The Lost Art of Retinal Drawing

The Lost Art of Retinal Drawing

By
Stephen R. Russell, MD
Dina J. Schrage Professor of Macular Degeneration
Director, Vitreoretinal Service
Clinical Director, Carver Center for Macular Degeneration
Department of Ophthalmology and Visual Sciences
The University of Iowa

LuAnn Dvorak, PhD, LPN
Faculty Lecturer
Department of Rhetoric
The University of Iowa

Foreword by Robert C. Watzke, MD

2013
Iowa Retinal Publishers
Iowa City, IA

Iowa Retinal Publishers

422 Butternut Lane
Iowa City, Iowa 52246

Library of Congress Cataloging-in-Publication Data

The lost art of retinal drawing / Stephen R. Russell, LuAnn Dvorak ; cover illustration by C. Neal Jepson ; back cover illustration by Kean Oh.

ISBN-13: 978-0615691978
ISBN-10: 0615691978

1. Art in Medicine 2. Medical illustration 3. Retina diseases

I. Title

RE14.R888 617.7092

THE LOST ART OF RETINAL DRAWING ISBN-13: 978-0615691978
ISBN-10: 0615691978

Printed in the United States of America.

Contents

Foreword

Ocular fundus drawing is almost as old as the ophthalmoscope. In 1863, Richard Liebreich published the first atlas of the human fundus, just twelve years after Hermann von Helmholtz introduced his ophthalmoscope. At first, drawings were made to introduce physicians to the features of the inner eye, both normal and diseased, and to assist in diagnosis. With better illumination and instrumentation, and particularly the dissemination of the handheld binocular indirect ophthalmoscope of Marc-Antoine Giraud-Teulon in 1861, the fundus could be examined in depth and breadth from the peripheral retina near to its limits at the ora serrata.

Following recognition by Jules Gonin that retinal tears caused retinal detachments and that the identification and closure of all tears resulted in a cure for this previously untreatable and common malady, the ophthalmoscope was elevated to a mandatory tool for preoperative and operative guidance. In Gonin's time, the ophthalmoscope was essential for meticulous examination and drawing of the fundus, which he insisted was fundamental to successful surgical results. In the words of Professor Marc Amsler, a former student of Gonin, "Gonin became angry each time a visitor acted as though he were interested only in the operation without caring at all for the work accomplished in the examining room. 'We must learn to do ophthalmoscopy,' one of the then German masters of ophthalmology said one day after leaving the examining room of Gonin."

While Gonin made his preoperative drawings and planned his surgery using a monocular indirect ophthalmoscope, the location of the retinal tears was estimated by an elaborate system of plotting the tears' meridian and distance from the corneo-scleral limbus prior to surgery. It was Charles Schepens who first combined binocular indirect ophthalmoscopy, through-the-lid scleral indentation, preoperative fundus drawing, and the use of scleral indentation at surgery to increase the success rate of detachment surgery from the approximately 60 percent in Gonin's time to the 90-plus percent reattachment rates of today.

Scleral indentation to bring the extreme peripheral retina into view was not invented by Schepens. But the use of this technique as an essential aid in examining the pre-equatorial retina in silhouette, while actively indenting the sclera through the upper and lower lid, was widely taught and advocated by Schepens. He also insisted on the value of a meticulous drawing of both fundi before surgery with use of such indentation on the sclera at surgery to locate and "prove" each possible tear. His technique of externally indenting the globe or "buckling the sclera" under each tear greatly reduced the postoperative limitations on ambulation, but his method of examination may have been his longer-lasting legacy and was fully as important a contribution.

When Edward Ferguson and I joined the Department of Ophthalmology faculty at Iowa in 1957 and 1958, respectively, we had the extreme good fortune and privilege of introducing Schepens's examination and surgical techniques. Both of us were convinced of their value and applied them when training ophthalmology residents and, later, retinal fellows to perform retinal reattachment surgery. We thought it essential that each resident gain use of indirect ophthalmoscopy and "through the lid" scleral indentation to construct a detailed drawing of the detached retina and the retina of its fellow eye. It was not enough to find each tear or hole and determine the boundaries between attached and detached retina. We believed that one should also recognize and draw normal features of the peripheral retina such as oral "teeth and bays," meridional folds, oral "pearls," granular tags, cystoid degenerations, and more. Only if they were drawn could we be sure that the resident had actually seen these features. "If it's not in the drawing, I don't know if you have seen it" was the mantra. And if you could not see normal features, how could you recognize abnormal ones?

So each resident or fellow on the Department's Retina Service was sequestered together in an examination room, the patient lay supine on an examining table, a chart was placed on the patient's chest, and the process of drawing the fundus features of a partially or completely detached retina was begun. This was usually done in the evening about two days prior to the scheduled surgery. Each color used had a different meaning: blue for detached retina, red for attached retina, green for vitreous features. The relationship of vitreous and retinal pathology was taught as well, but through examination of the vitreous cavity by slit lamp biomicroscopy and contact lens.

During the drawing procedure, residents had to learn to press gently on the lid to indent the sclera with an "indenter" while urging the patient to look in a certain direction in spite of an uncomfortably bright light. Since the image in indirect ophthalmoscopy is reversed, all features viewed had to be reoriented in the drawing to correspond with their actual location. After the first eye, the fellow eye was drawn too, since tears, subclinical detachments, and other pathological symmetry might be found there. Both resident and patient knew that if tears were missed, the operation would fail. There were no shortcuts.

Many residents insisted, vocally, that they had no artistic ability. Others would add depth and realism to their drawings by shading and detailing. But when the staff physician examined the patient the next day, the resident's examination was judged only by the correlation of anatomic details in the eye to what was depicted in the drawing. How many times was a resident asked, "Is that a granular tag or a tear at ten o'clock, right there where you have drawn it as all blue?" or "Did you see that enclosed ora bay at three o'clock?" or "Is that a peripheral hole?" After the session, the resident reexamined the patient until the staff was satisfied.

The drawings were used when the patients returned at each post-operative visit and were consulted again for research reports. And the drawings were used to ensure that each examiner was competent in the use of the indirect ophthalmoscope with scleral indentation before doing closely supervised surgery. The Retina Service staff at the University of Iowa achieved a successful surgical result in approximately 95 percent of cases and trained 188 residents and 32 retina fellows over that period. These doctors then left and applied those skills to the care and cure of numerous patients elsewhere.

The Lost Art of Retinal Drawing contains just a small sample of the drawings of all patients admitted and operated on for detached retinas by those University of Iowa residents, fellows, and staff from 1958 to nearly 1990. These drawings were considered so valuable that they were kept separate from the hospital charts; indeed, they were stored so well that only by good fortune were they recovered, and recovered intact and readable after many years of storage in an obscure spot.

The drawings handsomely assembled here by Steve Russell and LuAnn Dvorak reflect a time when retinal surgery was changing rapidly but was still based on fundamental principles as taught by the early masters. The fundus drawings of this period attest to the devotion and skill of the physicians and to the trust of the patients in their care. The drawings in this important book are a legacy of a time that is the foundation for the accomplishments of today.

Robert C. Watzke, MD
July 2011

Acknowledgments

We are tremendously indebted to numerous people for their unselfish contributions to this work. First we thank the artists of the retinal drawings. We appreciate their artistic contributions: the images, and we thank many from the group for their literary contributions including the touching and witty remembrances about times gone by. Special recognition goes to Robert "Bob" Watzke as an artist, mentor of artists, and contributor of the Foreword. This book is dedicated to Dr. Watzke, Dr. Edward "Ed" Ferguson, and the early inspirational faculty and trainees who established an observant culture at the University of Iowa. Dr. Tom Burton, who shared prominent mentorship during the retinal drawing era as reflected by the insightful comments included from several artists, also gets our thanks.

Second, we want to recognize the circumstances surrounding the art that had been lost and its transformation into book form. We are grateful to Connie (Fountain) Hinz and Kathy Burkle for their efforts in tracking down possible locations and gaining offsite access to document storage. The eventual discovery of the art and the fortuitous meeting of the authors, with our differing and complementary skills and interests, began the path to publication. We both are grateful for this spontaneous and productive event.

We also thank those whose support we received in the organization of historical and artistic elements. Patricia "Trish" Duffel, librarian and so much more, provided expertise in almost all areas – from computer programming to image sizing to publication information to finding resources. Her assistant, Rita Gallo, laboriously scanned and re-scanned the fundus drawings and performed de-identification, a requirement of our Institutional Review Board (IRB; Human Subjects Committee). Permission for use of the medical records, i.e. fundus drawings, was given by our IRB.

Additional support came from many others in and around the Department of Ophthalmology. Kathryn Sherman, for one, helped by answering questions about practical issues and by providing an additional perspective on artistic matters. Deirdre Davis offered to us her knowledge of art movements and her unfettered excitement about the retinal drawings as works of art. Karen Kelly and the chart control crew helped us access patient records, which were needed to identify or confirm artists and corroborate diagnoses.

As the project progressed, many individuals from the University of Iowa community, past and present, assisted us. Dr. Andy Doan provided important guidance through sharing his ideas and expertise with online publishing. The organizers of and participants at the Examined Life Conference gave to us an audience, some warm, witty, fulfilling feedback on our ideas, and a vehicle for public visibility at the conference and beyond. And several art historians at Iowa gave to us their time, expressed interest in the drawings, and (through their art versus craft arguments) motivated us.

A few key Iowa Ophthalmology department alumni critically read our manuscript, made suggestions, and offered encouragement. We appreciate the feedback from our readers Kathy Lyons (an alum from the Department of English), and Drs. Bob Baller, Neal Jepson, Tom Weingeist, Steve Wolken, and especially Stan Thompson who offered copious comments on both content and editing.

Finally, we wish to thank our daughters Carolyn "Dazi" Russell and Bethany Ryan. Steve's daughter Dazi contributed her art appreciation and knowledge from college art history studies to make the initial selection of representative fundus images. She selected 900 of the approximately 12,000 retinal drawings for consideration when she and Cheyanne Lester digitized all of the images, which had been consigned to post-digital destruction. LuAnn's daughter Bethany used her college graphic design studies as she created the initial graphic layout that served as the basis for the book proposal. Bethany's design also became the basis for the final chapter title pages taken up by our graphic designer Robyn Hepker. Many thanks to each of these women for their ideas and efforts!

Steve Russell and LuAnn Dvorak, 2013

Preface: *Finding the Lost Art*

I was one of the last retinal fellows at the University of Iowa to do formal retinal drawings. The purpose and opportunity for these drawings began to fade in the late 1980s as administrators and insurance carriers largely proscribed preoperative admissions.

When I returned to the University of Iowa in 1997, I realized the value and significance of the departmental artistic legacy. Dr. Lee Alward's book *Color Atlas of Gonioscopy* had just been republished by the American Academy of Ophthalmology and has become a classic instruction and reference for hundreds of ophthalmologists. The core of the book and its inspiration were fabulous drawings made several decades earlier by Lee Allen, a departmental photographer and artist who, prior to his position in the department, had developed his art technique and style under artist Grant Wood. The iris and iridocorneal angle drawings were nothing short of spectacular.

I also recognized, upon my return, that the departmental artistic legacy also rested on those retinal drawings that, I realized by then, had been discontinued and had faded from training programs and practices with no more than a whimper. Yes, a few old-timers in the retina field bemoaned that newer colleagues were not as observant, but the world had changed over the years. The rise of vitrectomy surgery, the near elimination of scleral buckling in some "progressive" programs, the discouragement of preoperative hospitalization, and the endless drive for "efficiency" in medicine were, when combined, the forces behind the discontinuation of the formal drawings.

So began my search for the drawings. I looked in the retina clinic, and in every storage area that I could identify. Everyone remembered the fabulous images; no one remembered where they were. After some months looking in the department, it was clear not only that preoperative retinal drawing had become a lost art form, but also that the retinal drawings themselves had likely been lost.

I then started detective work as any clinician would. I talked to the oldest and most knowledgeable nurses in the clinic. I called retired administrators. We identified all the known storage facilities in which materials from the department had been put. I was on a scavenger hunt. Our clinical research coordinator and I began visiting each of the University's many storage facilities that contained departmental valuables. I combed through old microscopes, records of old research studies, echography images, emeritus faculty 35mm slides, examination equipment, and research supplies.

Many of the storage areas looked like the last scene in *Raiders of the Lost Ark*, with rows of storage shelves stretching on and on. I was convinced that twenty boxes of retinal drawings, a.k.a. medical records, could not disappear so easily, but after exhausting all the known locations on our records, I thought the drawings might be lost to posterity. But I always knew that they had to be somewhere.

Eventually some archeologist would discover the cache of records that were, I imagined, hidden where I could not find them, camouflaged among masses of stored "valuables." Finally, I had exhausted every facility, every lead, and my attention span.

In 2006, one of the employees from a storage facility in Cedar Rapids called our department, asking if they could destroy some old drawings. I immediately requested that the drawings be sent home to the department. Of course the departmental administration did not understand their value and had no place to put nineteen boxes, so in the short term we stacked them in the retinal fellows corridor and my office, and we begged and borrowed space from our clinical trials coordinator area. We realized soon, however, that we did not have space in the long term to store the more than twelve thousand images that had been found.

Despite the space issue, the retinal drawings lay fallow for three years until the summer of 2009, when my daughter Carolyn "Dazi" Russell, who had done an art history minor, was working in the department. Our good will for internal storage had at that point run out, and based on the scale of the project, we decided to digitize all the images so the information would not be lost. However, after viewing a few of the images, I decided to save those most artistically, stylistically, or historically varied.

Over eight months, the pile of choice retinal drawings began to grow. When all the images had been scanned and about nine hundred of the original images remained, the project went into hibernation for another year, this time due to lack of resources.

Sitting down with the retinal drawings was a breathtaking experience. Knowing that each drawing took from thirty minutes to three hours, and required full cooperation of the patient and effort by the ophthalmologist, made the drawings even more remarkable. Each drawing was a cooperative product, as this project has come to be.

In September 2010 LuAnn Dvorak joined the project as the editor and coauthor. After her numerous trials to disabuse me of my original book concept (a retinal drawing with comparative formal art on the facing page with commentary), we decided to select a subset of retinal drawings to incorporate. The included images were those most varied in artistic style and historical merit, or the most interesting and varied imagery. On September 27, 2010, we laid out the images to view them simultaneously and make our choice (Figure 1). After our initial selections, we invited the retinal faculty, fellows, and residents to help in the decision-making. From the beginning until now, we have reduced the number of retinal drawings from 12,000 to 131.

Around the time of subset selection day, LuAnn visited a number of art historians who were generous with their time and information and excited about the project's historical significance, but they urged restraint with our terminology. These art historians explained that retinal drawings are not "art" since the intention at their genesis was not to produce art, but to represent a "craft" process of documentation.

As we see it, art and beauty have long been recognized subjective appellations. If the crafting of medicine to the individual can be considered an art, then so can this.

Stephen Russell, MD

Figure 1.
September 27, 2010. Making the final selection of images.

Introduction

"The important thing is not the sketch itself but the process of making it."

G. F. Hilton, E. B. McLean, and D. A. Brinton
Retinal Detachment: Principles and Practice

As a diagnostic tool and record of observations and analyses in an ophthalmologist's examination room, a drawing of the ocular fundus—the interior surface of the eye, which includes the retina, optic disc, macula, fovea, and posterior pole—communicates information to the physician or examiner who is making the drawing and to other physicians who will assist in treatment. At the heart of Hilton, McLean, and Brinton's statement lies the revelation that in making a retinal drawing the physician has educated himself or herself on the patient's condition to a degree beyond casual observance. Artistically speaking, the retinal drawing, which represents the physician's perception and interpretation of the image, is "the important thing" and the main subject of this book.

The title of this effort, *The Lost Art of Retinal Drawing*, contains several word plays. First, it conveys two meanings of the word *lost.* Because the production of retinal drawings is no longer commonplace, the art of drawing has been lost. And because the images shown in this book were misplaced for more than nine years, they too were lost. Second, the wording in the title is somewhat misleading because the drawings contain imagery of multiple layers and structures, more than the retina alone. Most ophthalmologists use the term "fundus drawings" in their communications and documentation to refer to illustrations of this kind, as the ocular fundus includes the optic nerve, retina, choroid, ora serrata, and more. However, in common parlance, the terms "retinal drawing" and "fundus drawing" are used interchangeably, and this tradition will be continued throughout this book.

To draw the fundus, an examiner wears on his or her head an apparatus called an indirect ophthalmoscope, which shines an illuminating light into the subject's dilated pupil. While holding a condensing lens in line with the retinal structure being observed, the examiner then draws with colored pencils various features of the retina and other fundus structures. To view the peripheral fundus (the area that is more than 90 degrees from the visual axis), the examiner indents the globe by pushing on the eyelid with a scleral depressor, which may be a metal paddle, a cotton-tipped applicator, or another smooth instrument.

The fundus is typically drawn on a sheet of card stock, slightly larger and wider than standard notebook paper. A formal fundus drawing has markings that indicate direction, resembling the face of a clock. Concentric circles symbolize the optic disc (the smallest and off-center circle), the equator, the ora serrata (the front-most extent of the retina), and the ciliary body. Because of the way the fundus is imaged, the representation seen by the examiner is upside down and backward, and the examiner does not have a continuous view of the fundus. Instead, each field of view is drawn and then the direction of gaze is repositioned to draw another field. Details of the process will be further discussed in chapter one, but it is important to note here that, due to alignment and orientation issues, this process typically took between thirty minutes and three hours, depending on the experience of the examiner and the cooperation of the patient.

The era of fundus drawing at the University of Iowa Hospitals and Clinics, from which the drawings in this volume came, lasted only a little more than thirty years: primarily from 1958 to 1988. For the first half of that interval, it was later discovered, a particular method of cataract extraction contributed a higher rate and number of the retinal detachments, which is evident here in the diagnoses that accompany each of the drawings. Altogether though, in the entire collection and throughout the entire thirty-year period, most of the drawings were rendered before retinal reattachment surgery to facilitate surgery by mapping the retina, retinal defects and the surrounding structures. The types of surgery for retinal detachment during the fundus drawing era included scleral buckling, which was reinvigorated in the late 1940s and early '50s, and vitrectomy, which was first reported in 1969 and developed rapidly in the early 1970s.

The 131 drawings in this volume represent the work of 75 artists and were selected neither randomly nor systematically from approximately 12,000 drawings in the collection at the University of Iowa. The selection process involved subjective criteria covering artistic and historical merit. In the end, the selected drawings had more artistic merit, were completed images rather than partial sketches, or offered a unique look or style. The selection process was not easy. However, the results of that process—as the readers of this volume will see—represent the impressive potentiality for ocular fundus representation.

The Lost Art of Retinal Drawing connects art, medicine, and history; showcases technique and talent; and honors an era. Starting from the beginning, chapter one "The Meticulous Examination of the Fundus" reveals the process of fundus drawing, where the practice took place, and what happened during the exam, including the influence of or the collaboration with other artists on the Retina Service. These influences appear to have led to stylistic similarities in drawings done by classmates or by teacher and student, and yet this chapter focuses most on the entire scene as well as on the players. Excerpts of how numerous artists remember their fundus drawing experiences and mentors accompany the drawings.

Stylistic similarities are evident in the images of individual artists and some pairs of classmates, and most remarkably across larger groups of artists, as depicted in chapter two "How Group Drawing Styles Changed over Time." The drawings featured in that chapter have been divided into three time periods to show how, in merely thirty years, the art of fundus drawing evolved from realistic depictions in muted colors to increasingly abstract, colorful expressions to whimsical figures and delineations before passing from existence.

Individual stylistic differences, too, are evident within the drawing collection. Although it seems obvious that individual artists would have their own way of drawing, just as they have their own way of writing their signatures or annotating the images, the variations can be striking. For example, in chapter three "Gradations of Shading," various techniques of shading are revealed in the form of complexity, sophistication, and artistic intent. And chapter four "The Representational Range of Fundus Abnormalities" shows that a close inspection of the drawings and accompanying text can reveal how much the same anatomical feature or pathological condition visually differs depending on who drew it. This chapter features drawings of giant tears, cysts, and various types of detachments.

"Limitations of the Color Convention and Use of Pseudo-Color," lovingly known as "the green chapter," centers on the different ways that transparency, membranes, and blood are portrayed in the drawings. Although green is not seen on a live retina, the images in chapter five all contain that color, and some of the images, specifically those found in the chapter's final image sections, go beyond even the standardized symbolic color chart in reflecting particular artists' styles and vision.

Finally, this look at the differences in the drawings culminates in chapter six "The Art of the Ora Serrated: The Complexities of the Hidden Edge of the Retina," with a look at the issue surrounding this book: whether the drawings are art, via the structure that surrounds the retina as it is drawn. The ora serrata actually is the front edge of the retina but is drawn as the middle concentric circle. Adjoining the pars plana, the ora serrata is difficult to visualize but, because of the association of certain types of retinal tears and deformities of the ora, may provide particularly useful landmarks when an examiner draws the fundus.

What is significant here is that how the ora serrata is drawn suggests artistic intent, as much as and possibly more than drawing any anatomical feature of the fundus, since the ora serrata seems at the mercy of artistic license. In truth, deviation from the norm is evident throughout this volume, especially because that norm changed dramatically during the era of fundus drawing. But the ora serrata is a feature forgotten in numerous drawings, scribbled in others, realistically depicted in some, and glorified in a few. Its presence and appearance in a fundus drawing reflects artistic intent, albeit secondary to medical documentation by necessity.

The extent of the stylistic similarities and differences in the drawings in this volume, and what the differences or deviations from the universal fundus chart "color code" might mean in practice, are more evident to ophthalmologists; however, the art itself is the focus of this book.

The Lost Art of Retinal Drawing may be the only place to see this type of art and not only because the drawings are no longer done, which of course they are not. A key reason that another collection of this size and variety does not exist relates to decisions elsewhere to place individual drawings inside patients' medical records. At the University of Iowa, however, the drawings were kept in separate files that were eventually recovered after an arduous process of intellectual excavation, physical searching, and expectant waiting. To identify and collect drawings at other institutions, thousands of records would have to be reviewed after securing lists of past surgeries and patient permissions. The resources involved in such a project would have been extensive and likely prohibitive. In addition, the medical records themselves are now in most cases being digitized and the hard copies destroyed as a consequence of the evolution of medical records—now electronic—and decisions of administrative personnel. Of course, with the destruction or reduction of those records comes the loss of original drawings.

In other words, this collection may represent the largest and most comprehensive collection of this particular type of relic. And although the artists whose drawings and words are featured in this volume were affiliated with only one institution, the compilation is a testament to the skills and talents of all artists who did formal fundus drawing. This collection is documentation of their lost art.

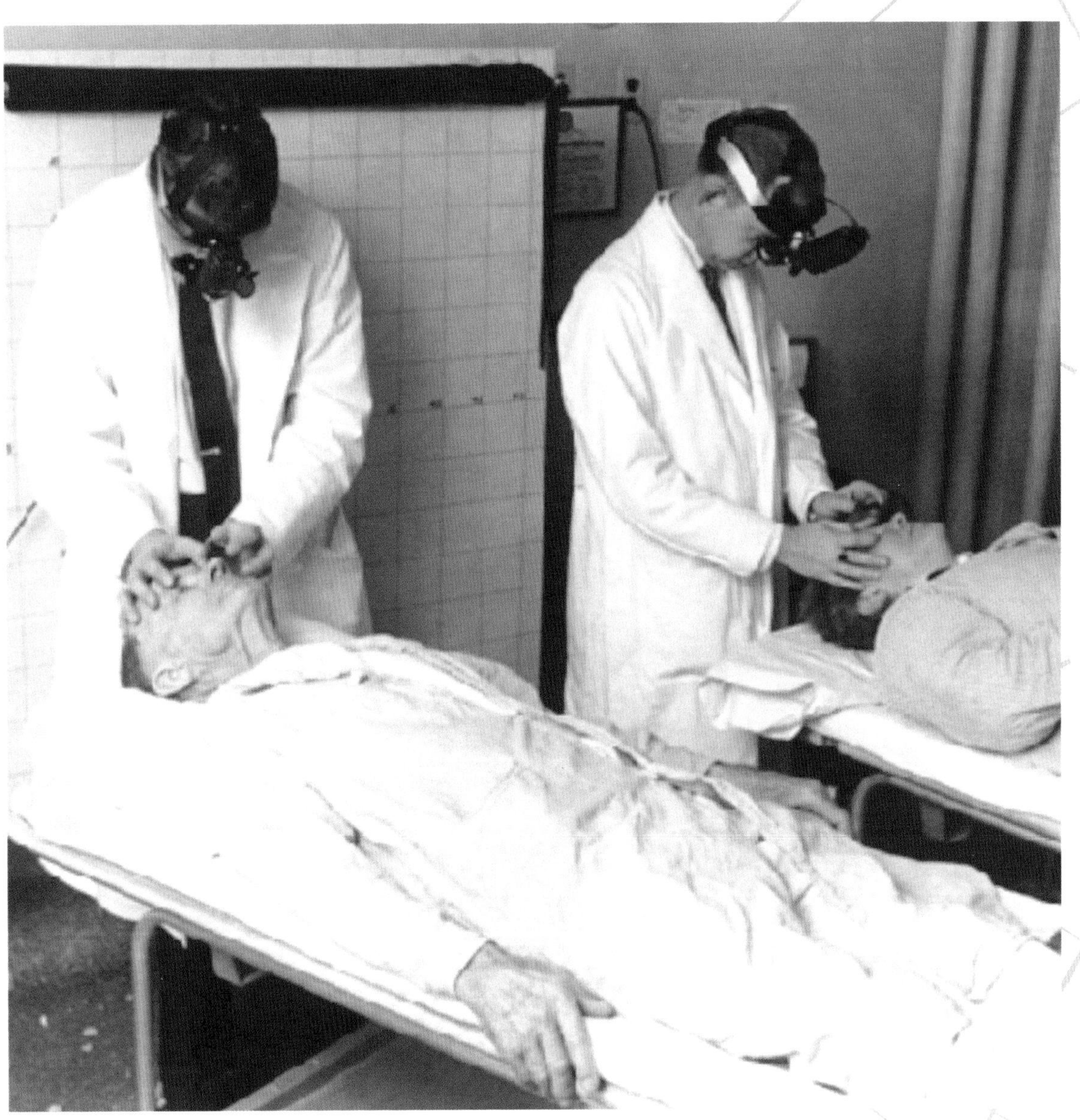

Retina faculty at work performing retinal examinations with indirect ophthalmoscopy, circa 1959 (Dr. Robert C. Watzke [left] and Dr. Edward C. Ferguson [right])

Chapter 1

"The Meticulous Examination of the Fundus"

If someone were to imagine a bullpen, a ballroom, a temple, and "a laser scene from *Star Wars*," the sounds, images, and emotions that enter the person's mind would greatly differ depending, of course, on which term the mind was focusing at any given moment.

A bullpen, for example, is a place where baseball pitchers, prisoners, or male bovines congregate. These places sound rough, gritty, mean. A ballroom is a place for dancing. There is music playing, a polished floor, gracefulness, and elegance. A temple is a place for worship, or at least a place of high moral and intellectual value. Such a place is free from the commotion and anxieties of daily life; it is a place for meditation. In a laser scene from Star Wars, there would be fast action, lights moving at different angles, contact, and precision. These four entities create such disparate images in a person's mind and have totally different connotations, and yet each of these terms or phrases—a bullpen, a ballroom, a temple, and a laser scene from *Star Wars*—has been used to describe the retina room at the University of Iowa Hospitals and Clinics (UIHC) where ocular fundus drawings were done from 1958 until 1988.

It is not difficult to understand how one setting can create such different impressions. As time elapsed, as treatments changed, as hospitalizations shortened, as drawing went from one big room to individual patient rooms, and as technology improved, the physicians in the Department of Ophthalmology—or, rather, the *artists*—experienced fundus drawing in new and various ways. Those artists—faculty, residents, and (starting in 1968, then restarting in 1973) retina fellows—were individuals with diverse backgrounds, with different types and amounts of training, and with various levels of artistic skill or talent. Who an artist is, it seems, has much to do with the metaphors that the artist uses to describe an environment or object he or she sees, and possibly has even more to do with how the artist, the examining artist in this case, represents on paper the eye that is seen with the eye.

"By some miracle [the patient] is attached still in [the] right eye"

Robert C. Watzke, MD (11/13/72)

"Residents on the Retina Service lived like bats. In the dark most of the day to keep dark-adapted."

Stanley Altman, MD (12/2/10)

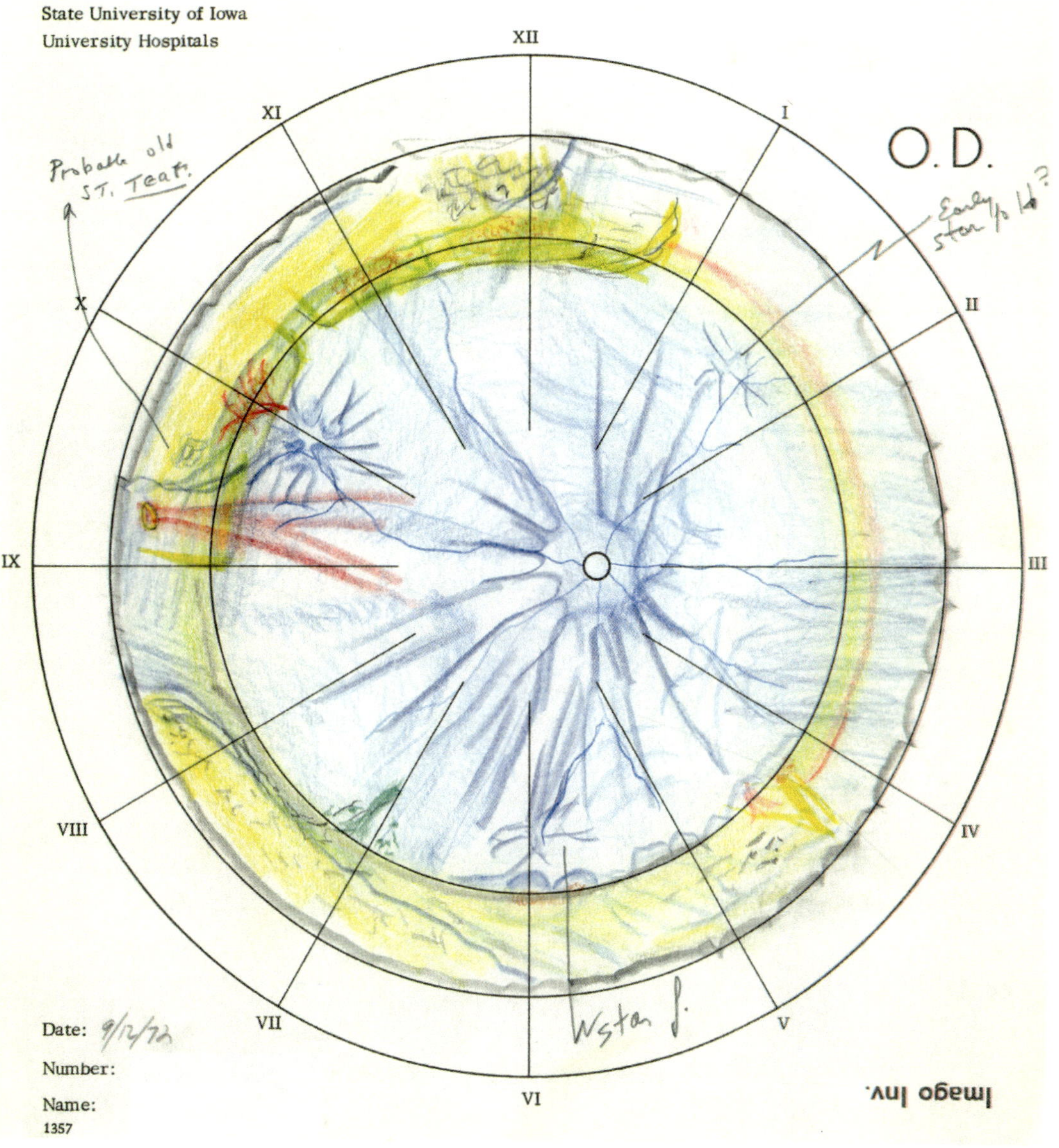

Artist Stanley B. Altman
September 12, 1972

Diagnosis:
Retinal detachment, right eye, and fifty-pound weight loss in ten months.

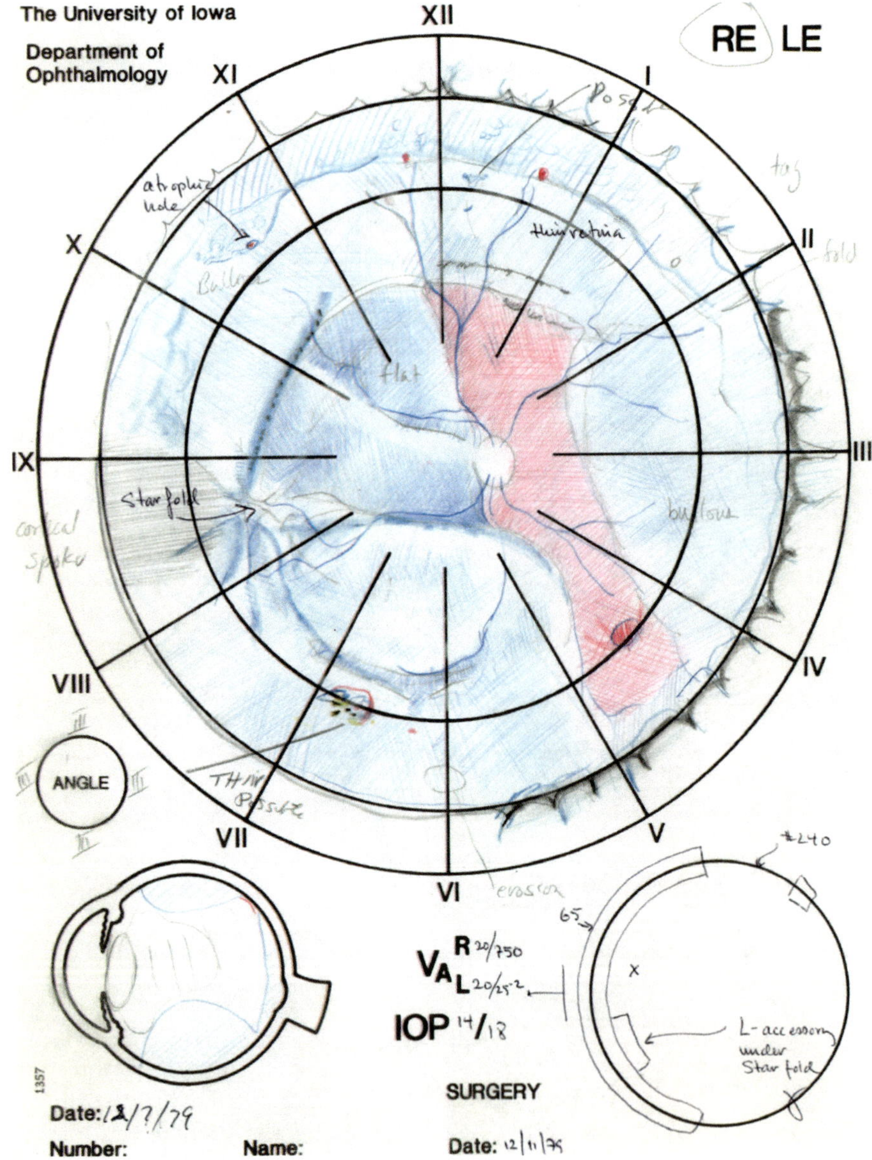

Artist
Manfred A. von Fricken
December 7, 1979

Diagnosis:
Almost total retinal detachment, right eye; very thin retina superonasally.

"Fundus drawing was a "tedious but valuable teaching chore. It sure did help us learn to use the indirect ophthalmoscope and required good patient cooperation."

Clyde Kitchen, MD (12/6/10)

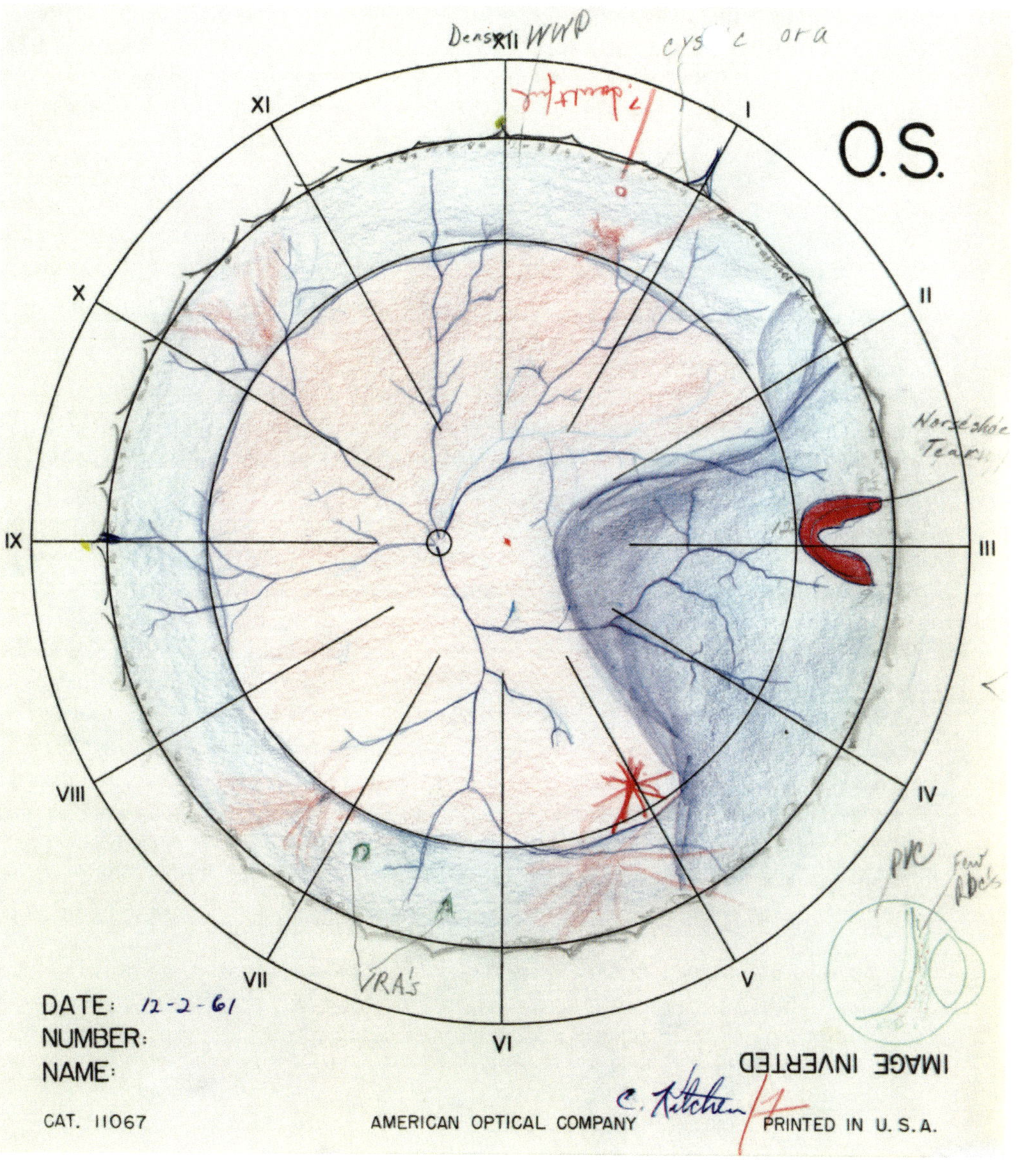

Artist Clyde K. Kitchen
December 2, 1961

Diagnosis:
Retinal detachment, left eye.

In most cases, the fundus drawings were done before surgery on patients who had retinal detachments (RD), where the retina separates from the tissue beneath it. Most often a tear or "break" in the retina, and the ensuing escape of fluid through it, led to the detachment. While mentioning these breaks that not only led to the detachment but also would need to be mended in surgery, Dr. Manfred A. von Fricken reflected about the practice of fundus drawing and the necessary time commitment:

> *We would examine the RD patients pre-op in the clinic on one of the stretchers and spend an hour or two drawing the retina to identify all of the breaks so that they could be addressed at the time of surgery. It certainly made us all excellent examiners and, at least in my case, a so-so artist. Over time the colors have faded, but the type of surgery done at that time was the very state of the "art." (12/8/10)*

The time spent examining the retina and surrounding structures, and the use of a new—in the 1950s—method of examination employing a light called the indirect ophthalmoscope, a handheld lens, and a technique where the sclera (the outer layer or "white" of the eye) would be "depressed" (pushed in), were major factors in the drawing process and actually what allowed it to happen, since the peripheral fundus cannot be visualized otherwise.

Dr. von Fricken further explained not only how the exam lends itself to the artwork and the artwork affects the exam, but also how the artist represents on paper what is seen:

> *The experience of making the detailed drawings helped us identify intra-operative problems and made us experts at scleral depression and seeing the inside of the eye with an indirect ophthalmoscope. It seemed to me that we could almost visualize and fix the retinal detachments having studied them for a long time, much as one would visualize any work of art, that the intimate knowledge of that person's retinal anatomy and features would project toward a successful operation. I always visualized what I would do intra-operatively and tried to make that happen. [It was a] very right-brain process of image processing and pattern recognition, which, I believe, helped me become a good surgeon. I would also practice and make postoperative notes of what went well and what didn't, often learning the most from the difficult cases.*

It is important to note that here Dr. von Fricken mentions his future, when he would "become a good surgeon." This is important because it is a reminder that these drawings were done in an academic setting where specialists would train novices but where, again, the backgrounds, training, and skills, both medical and artistic, often differed among the players.

The Discipline

Two major players on the scene at Iowa who had shared experience, similar vision, and a similar style of drawing in the early years, but different styles later and different styles of teaching fundus drawing were Dr. Robert C. Watzke and Dr. Thomas C. Burton. Dr. Watzke arrived in Iowa in 1958, just after Dr. Edward C. Ferguson arrived. The two had been in Boston, where they learned the methods of Belgian-born Dr. Charles L. Schepens, who had developed the modern indirect ophthalmoscope. Drs. Ferguson and Watzke brought with them to Iowa not only the indirect ophthalmoscope, but also scleral buckling surgery and retinal drawing.

Ten years later, Dr. Burton was Dr. Watzke's fellow—actually the first fellow in Retina at UIHC. During the research for this book, many of the artists remarked about their "teachers," with most of the comments singling out Dr. Watzke, Dr. Burton, or both. The consensus was that Drs. Watzke and Burton both greatly influenced how residents and fellows learned to treat retinal disorders or pathology through, as Dr. Burton called it, "becoming familiar with the eye" (12/14/10).

Dr. Watzke ("RCW" to many; "Bob" to some) went so far beyond the call of duty, as he cared for patients and educated residents, that he should be considered a saint, according to Dr. Robert S. Baller. In his nominating speech for Dr. Watzke's pseudo-canonization, Dr. Baller explained:

> *When I came to Iowa (knowing immediately it was my home and my fellow residents became my brothers and sisters), the anatomical cure rate for retinal detachment was 50 percent, and there had just been a real breakthrough in Charles Schepens's binocular indirect ophthalmoscopy with indentation and resultant immediate finding of new retinal pathology, which accompanied it. The cure rate went up to 80 percent plus. The fact was of course that you could see 80 percent of the retina (out to about the equator), but 80 percent of the etiologic detachment-causing pathology was anterior to the equator and optics precluded any exam there unless the retina sagged so much or the patient was indented.*
>
> *When I got to Iowa, then, the only people who had been initiated into the secret society of peripheral retinal pathology, indentation, and that other crucial part of peripheral retinal, were Bob and the few residents already on the Retina Service. We presented pre-op patients to Dr. Braley (for whom I have great love and affection, as I do all my mentors at Iowa). When patients were presented by Bob or the retina fellow, Dr. Braley and others would come forward pro forma and look with the direct ophthalmoscope, of course not seeing anything but the detachment, and [yet] trusting completely that Bob knew his business. The wonderment of seeing a person with a skill set that Dr. Braley trusted so much inspired all the rest of us to admire and trust Bob as the "go to" guy.*

"I remember my first retina rotation—when junior residents learned of proper examination techniques and the art and judgment involved in retinal detachment surgery ... severely trying the patience of Doctor Bob Watzke as he taught and 'assisted' our all-thumbs, sweaty-brow, 'first, do no harm!' scleral buckling attempts."

David E. Townes, MD (12/07/10)

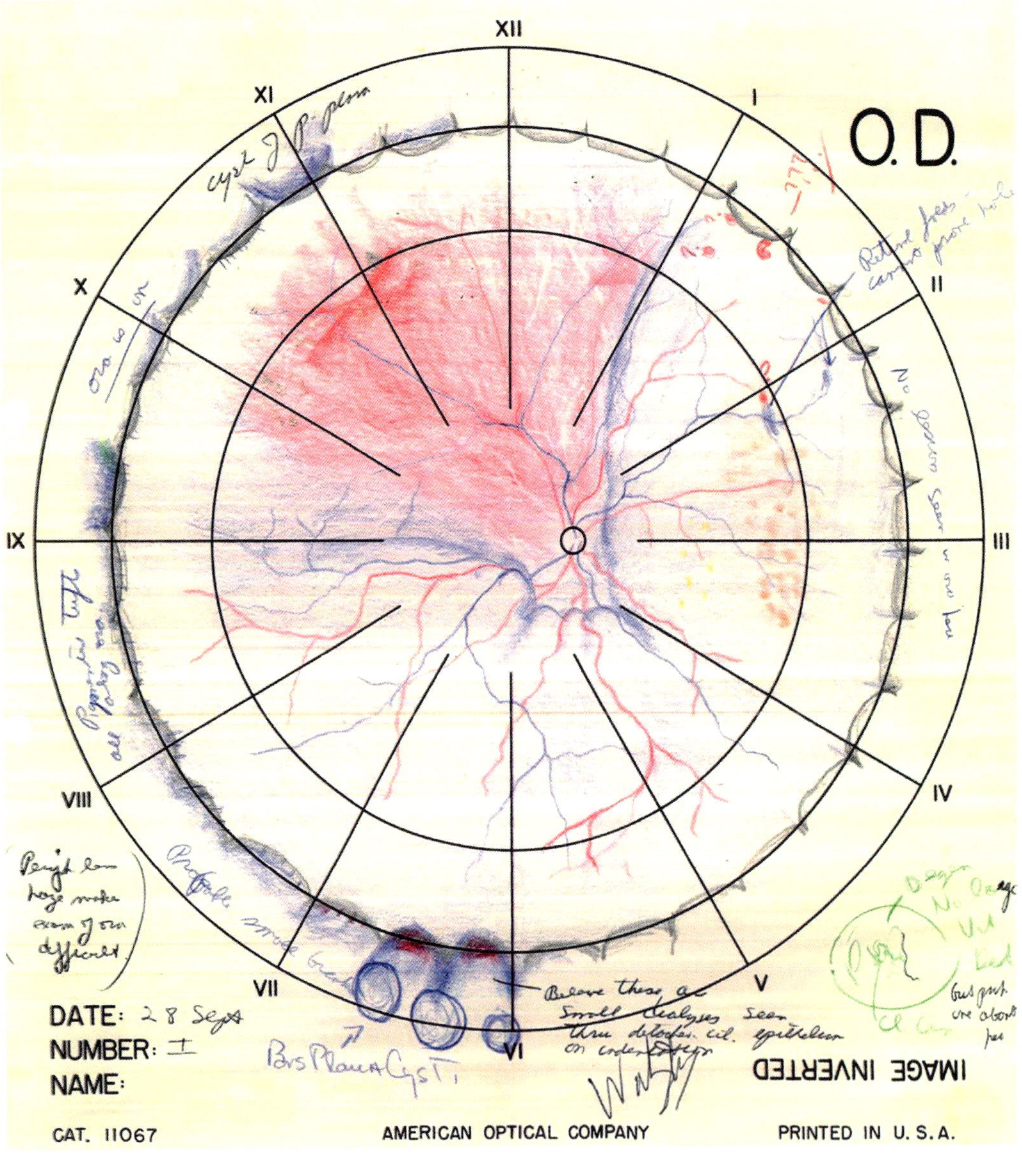

Artist Robert C. Watzke
September 28, 1959

Diagnosis:
Retinal detachment,
right eye.

"Patients would get accustomed to that brightness. … It was uncomfortable for them to get started" but after a time they "would be less light sensitive."

Thomas C. Burton, MD (12/14/10)

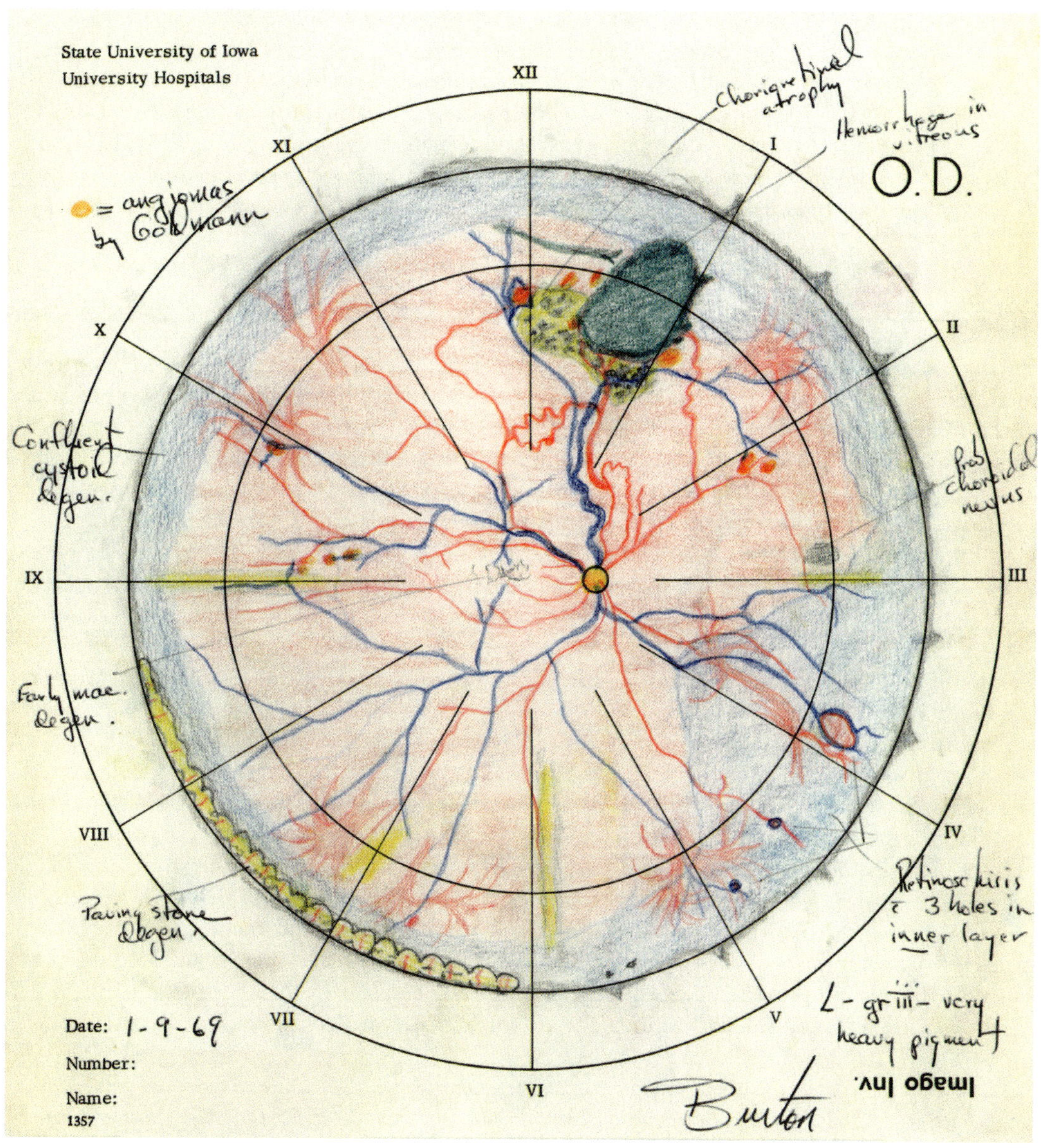

Artist Thomas C. Burton
January 9, 1969

Diagnosis:
Large angioma of retina, right eye, or Angiomatosis retinae (tumor/lesion), right eye.

> *According to legend, Bob learned to examine the peripheral retina by going home after work, dilating his wife Lonnie's eyes, and gaining those skills (So important Lonnie, of course, gets sainthood, too). Bob was known as a Renaissance man, having a Zeiss photocoagulator, a ruby laser (both with direct optics), [and] secret knowledge of the peripheral retina that he shared almost Zen like to his followers (residents), rebuilding and restoring a Model A (or T) in his garage, and being a wonderful person and role model of patience and knowledge. His patience in teaching us qualifies him for sainthood, both in the exam room and the OR. (12/27/10)*

Other artists praised Dr. Watzke and his influence on their educations and subsequent careers, but Dr. Baller was the most enthusiastic.

During the years that fundus drawing was done at UIHC, the Department of Ophthalmology included several faculty members who, in communications for this book, received praise for their patience and guidance from former residents and fellows. The list of those men or mentors includes Drs. Alson E. Braley, Thomas C. Burton, Richard F. Dreyer, Edward C. Ferguson, James C. Folk, Vernon M. Hermsen, William B. Snyder, and Thomas A. Weingeist, with the artists obviously praising their own mentors, such as when Dr. T. Kim Krummenacher said, "Thanks to Dr. Weingeist" for the surgical training and "Dr. Folk bears prominent mention" (6/6/11).

Of the individuals named on that list, however, only Dr. Burton was mentioned as much as Dr. Watzke, with both receiving credit for the considerable influence they had in relation to fundus drawing and its effect on surgeries. It must be mentioned that Dr. Burton was a fellow and then on the faculty at Iowa for a total of twelve years during the fundus drawing era, and Dr. Watzke was on the faculty for twenty-six.

Dr. Burton said he "learned how to do the basics [of fundus drawing] from Bob Watzke" (12/14/10) and then taught the practice as well. In the "ballroom," Dr. Burton reminisced, there was a lot of conversation but not among the patients, and very little privacy. Inside that retina room were faculty members, residents, fellows, medical students, and three or four cots or stretchers, each bearing one patient. "It was quite a zoo in there," he said, adding yet another metaphor to the mix.

The zoo had its benefits, both academic and practical. One benefit Dr. Burton mentioned was that there were "multiple doctors looking at a patient with a particularly interesting lesion that everyone could see" since the other doctors were only "three feet away." The doctors would see new diagnoses and rare pathology, hear the abnormalities discussed, and learn. This academic setting affected more than just the doctors, however. Dr. Bruce E. Herron noted:

> *I know that it was somewhat demanding on the patient. This certainly took more time than just examining the retina. Patients would not likely be aware that this required more*

"Usually we didn't have time to complete the drawings and would have to come back at night to finish them."

Paul O. Sanderson, MD (12/21/10)

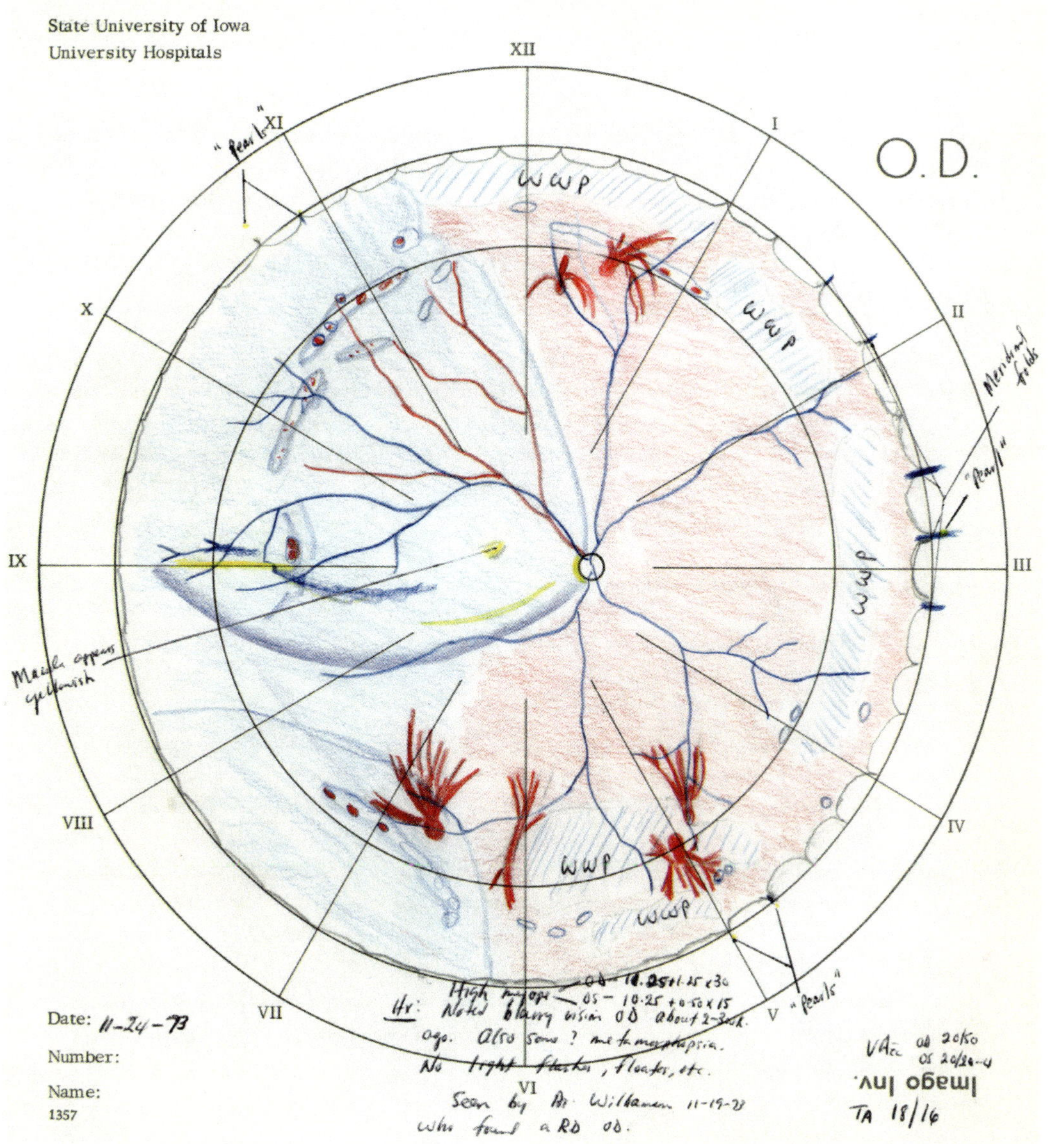

Artist Paul O. Sanderson
November 24, 1973

Diagnosis:
Retinal detachment, right eye; extensive lattice degeneration.

> *time and cooperation on their part, but would want to do everything necessary to restore their vision. Perhaps they would know that things may take longer in a "teaching hospital" but are willing to endure this for quality care. (2/26/11)*

A teaching hospital provides care to patients and clinical education to health professionals. Often additional time for examinations is necessary, as Dr. Herron states, for the medical student, student nurse, resident, or fellow to learn clinical procedures and disease processes.

In ophthalmology, drawing the fundus forced residents and fellows to find minute details and represent them on the fundus chart before a patient's surgery, with the faculty checking for accuracy along the way. Dr. von Fricken explained:

> *Fundus drawings were de rigueur at the time and this was true for all of the outstanding retina programs. Tom Burton would make it a point to take one of the drawings and actually "count" the teeth and bays of the ora serrata, in effect editing the fellow's drawing. It was a part of the culture of the meticulous examination of the fundus and the drawing of all of the vessels, breaks, vitreo-retinal abnormalities, ora bays and teeth, the vitreous base, and the extent of the retinal detachments. (12/8/10)*

Teeth counting and locating previously undetected small tears were the two most shared memories about faculty members reviewing the drawings and demonstrating their advanced skills. Dr. Peter R. Pavan mentioned one of those undetected tears when he elaborated on Dr. Burton's influence:

> *Retinal drawing was taught primarily by Dr. Thomas Burton when I was a fellow. He was very exacting. He would demand we count the number of ora bays on the nasal side and accurately reflect this number in our drawings. He also had 20/10 vision without correction. That and his experience allowed him to see holes I could not. I remember the day he said I did an excellent drawing, especially my accurate depiction of the retinal vessels. He then went on to say there was only one thing wrong. I had missed the retinal break responsible for the detachment. And this was after I had spent over an hour examining the patient and making the drawing. He then drew the tear in the V created by a bifurcating nasal retinal vessel I had accurately depicted. When I reexamined the patient, the tear was exactly where he had added it to the drawing. And, of course, when we buckled that tear and only that tear, the detachment was cured. (12/22/10)*

"While I seldom ever saw beyond the posterior pole, I had to draw it well thanks to Bob Watzke and Tom Burton. We had to make pretty pictures before they would let us do surgery. I think the retina drawing did more background for aesthetic eyelid surgery than the rest of the retina rotation."

Richard Anderson, M.D.

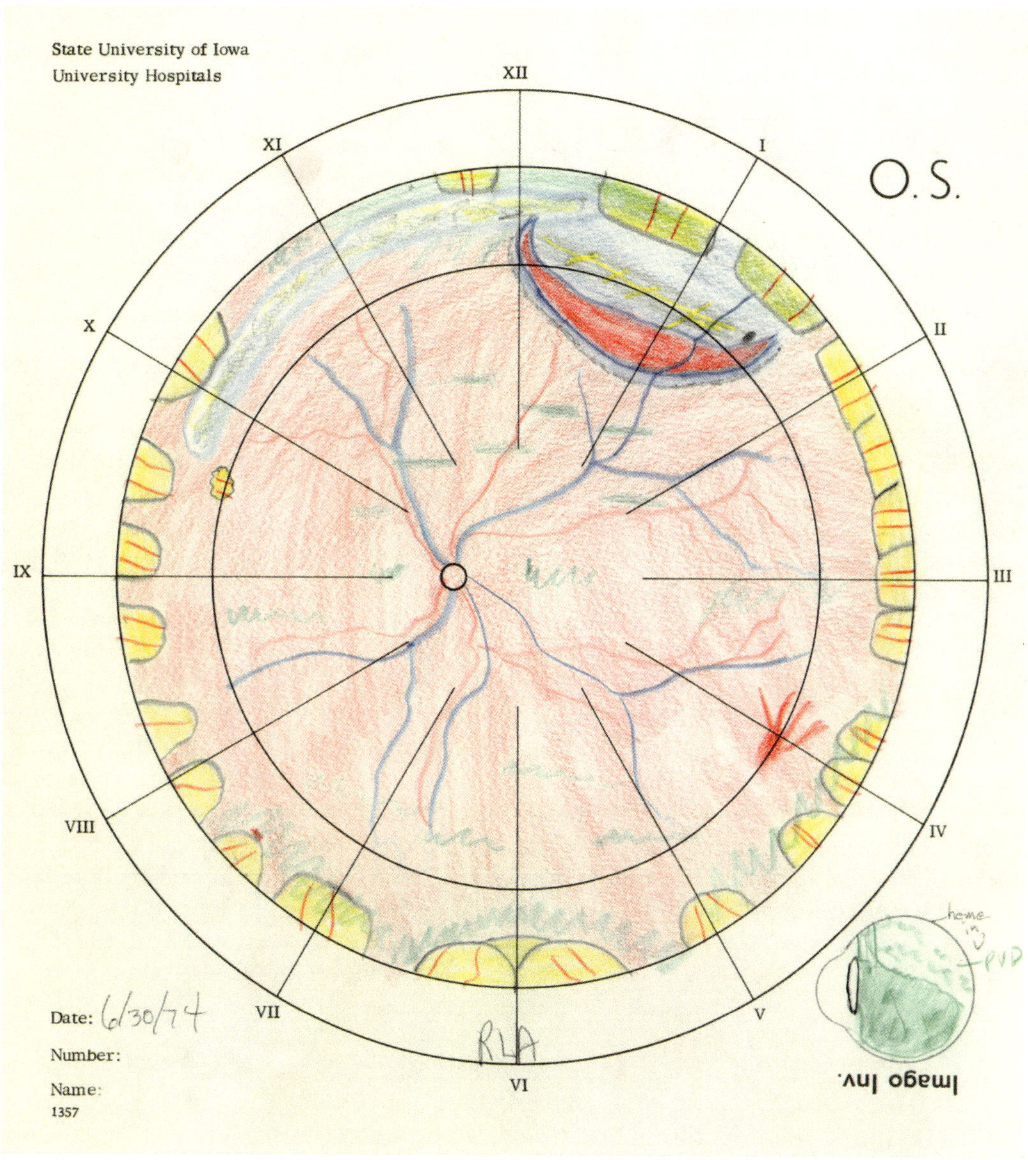

Artist Richard L. Anderson
June 30, 1974

Diagnosis:
Retinal Detachment, left eye, and Lattice Degeneration

Obviously Dr. Burton could have pointed to the tear right away, but he chose to use a less-direct teaching method, within reason. Another example of this particular style of teaching how to find an elusive tear came from Dr. Steven J. Vermillion, who said,

> *The best thing I learned from Dr. Burton was how to see fish in muddy water. He would look into an area of clouded blood that you had been looking at for an hour and tell you there was a hole there. You would look for another half hour and then he would come over, define the general area to look in, and tell you to just keep staring at this cloud of red haze. Suddenly, in the midst of that cloud, a microscopic hole would magically appear. It had been there all along, but your brain could not perceive it. That was the most marvelous thing I learned. (10/10/11)*

Words like *magically* and *marvelous* capture the essence of a successful teacher-student relationship and its potential benefit to the patients in this context.

The importance of fundus drawing for fellows and residents during the fundus drawing era must be underscored in red for two reasons. First, when looking at the retina, the ophthalmologist sees red—various shades of red, that is, as the other colors (blue for detachment and green for vitreous hemorrhage) merely represent pathology on the drawings. A second reason the importance of fundus drawing for fellows and residents must be underscored in red is the amount of knowledge these physicians gained. "Retina was the one service I learned the most on," former UIHC resident Dr. Paul O. Sanderson stated (12/21/10), but it was a sentiment echoed by many of the artists who did the retinal drawings in this book. Meticulous examination with a meticulous drawing had multiple benefits.

"I do not demand my fellows make these classic drawings," Dr. Pavan said, referring to his position as professor in vitreoretinal surgery and to current practices (12/22/10). "But like Tom Burton, I delight when I find tears they have missed after examining the patient. Of course, the fellows really hate it that an old decrepit [sic] codger like myself can show them up. Perhaps retinal drawing is no longer taught with the rigor of old but in the most important way the tradition of teacher to student continues." And that tradition continues in part due to the training that teachers of today received.

"A lot has been lost," Dr. Burton lamented while discussing the teaching practices, the benefits of doing the drawings, and the demise of a practice done back when doctors had "the luxury of admitting people," as Dr. Burton said (12/14/10). This allowed the doctors' time to complete the drawings by either staying late at the hospital or returning in the evening, after dinner, when the clinic was quiet.

In comments similar to those of other former residents, Dr. Sanderson said that he was "very concerned to be precise and accurate" while drawing, and like many of the other former residents and fellows, he told how Dr. Burton corrected the drawings during the training and sometimes

"In those days we drew lots of patients, and it was a busy service. Doing a drawing required around an hour of indenting and sketching all the landmarks. Those poor patients. Tom Burton or Watzke would check everyone's drawings always finding at least some detail you missed, taking only a few minutes."

Chase P. Hunter, MD (11/16/10)

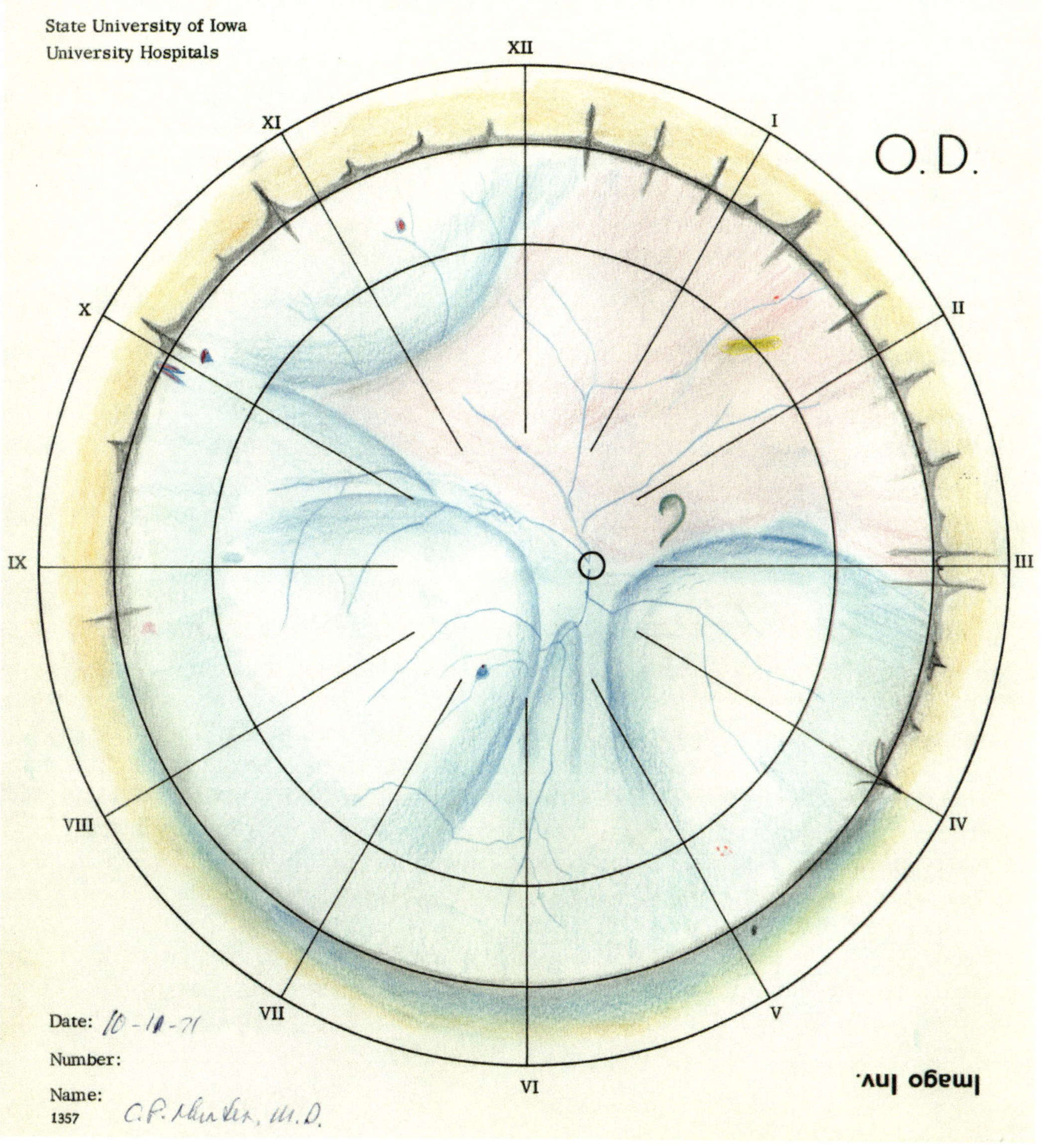

Artist Chase P. Hunter
October 11, 1971

Diagnosis:
Retinal detachment, right eye.

ripped up the drawings when there was an egregious error. But Dr. Burton himself added a couple of stories showing that an error on a drawing was not the only reason that, in general, he was seen as the disciplinarian who expected and directed, while Dr. Watzke was seen as warmly guiding his disciples.

One of Dr. Burton's stories concerns a resident who did not do his drawing in the evening, but showed up the next day for surgery. "He fully expected to do it," Dr. Burton explained—still astonished and perturbed after all this time—but he had said no. In another story, he told of a resident who "would use a graphite pencil to do a drawing and then take it home and use the colored pencils there. The rule was 'no elaboration at home' so I walked up to him and shredded it." Later in the conversation containing these stories, Dr. Burton added, "I must have been a hard person to get along with."

"In Tom's defense," Dr. Pavan said, "he may have been difficult because he had a difficult job."

> *Tom had to watch novices do retina surgery. Not an easy thing, trust me. He was very generous because I never knew him to take over a case, no matter how bad the resident or fellow was performing. He had real courage. ... Now that I have to watch my fellows operate, I truly understand Tom's patience with us. Those graduates of his who never had to teach a fellow or resident in the operating room may never fully understand all that Tom did for them. (12/24/10)*

Dr. Watzke echoed Dr. Pavan's emphasis on the importance of not only the drawings but also the academics. After Dr. Watzke had read some of the comments about drawing the retina as a map for surgery, he calmly corrected the record by adding, "I thought that the main purpose of drawing was to teach the use of the indirect and the need for meticulous examination" (4/26/11).

Dr. James G. Diamond seconded this idea concerning education as purpose. While comparing drawings done in the era of formal fundus drawing and the smaller, quicker, less ornate ones done since the early 1990s, he said, "While not an issue at Iowa when I was present [in the 1970s], drawings are reimbursed by insurance companies at the same rate as a simple wide-angle fundus photograph is today. A monetary explanation is more prevalent in a private setting while an educational gain is the primary stimulus to draw in a teaching institution" (5/23/11).

Of course the underlying reason retinal drawings were needed in the first place was the presence of a retinal detachment or a few other conditions. Most often a tear in the retina, and the ensuing escape of fluid through it, led to the detachment. When the cell structure became disturbed, vision did as well.

"I have been bemoaning the diminishment of fundus drawing over the last 10 years. Even sketches to show the location of pathology are done less and less commonly. ... I have two main memories of fundus drawing at U of I. (1) The then retina faculty Dr. Tom Burton insisted on us working on the drawing until it was accurate according to his view. (2) The amazing tolerance of Iowa patients for putting up with one to two hours of having someone stare into their eye with a bright light while probing over the eyelids in order to better see the pathology."

Tom Stevens, MD (12/7/10)

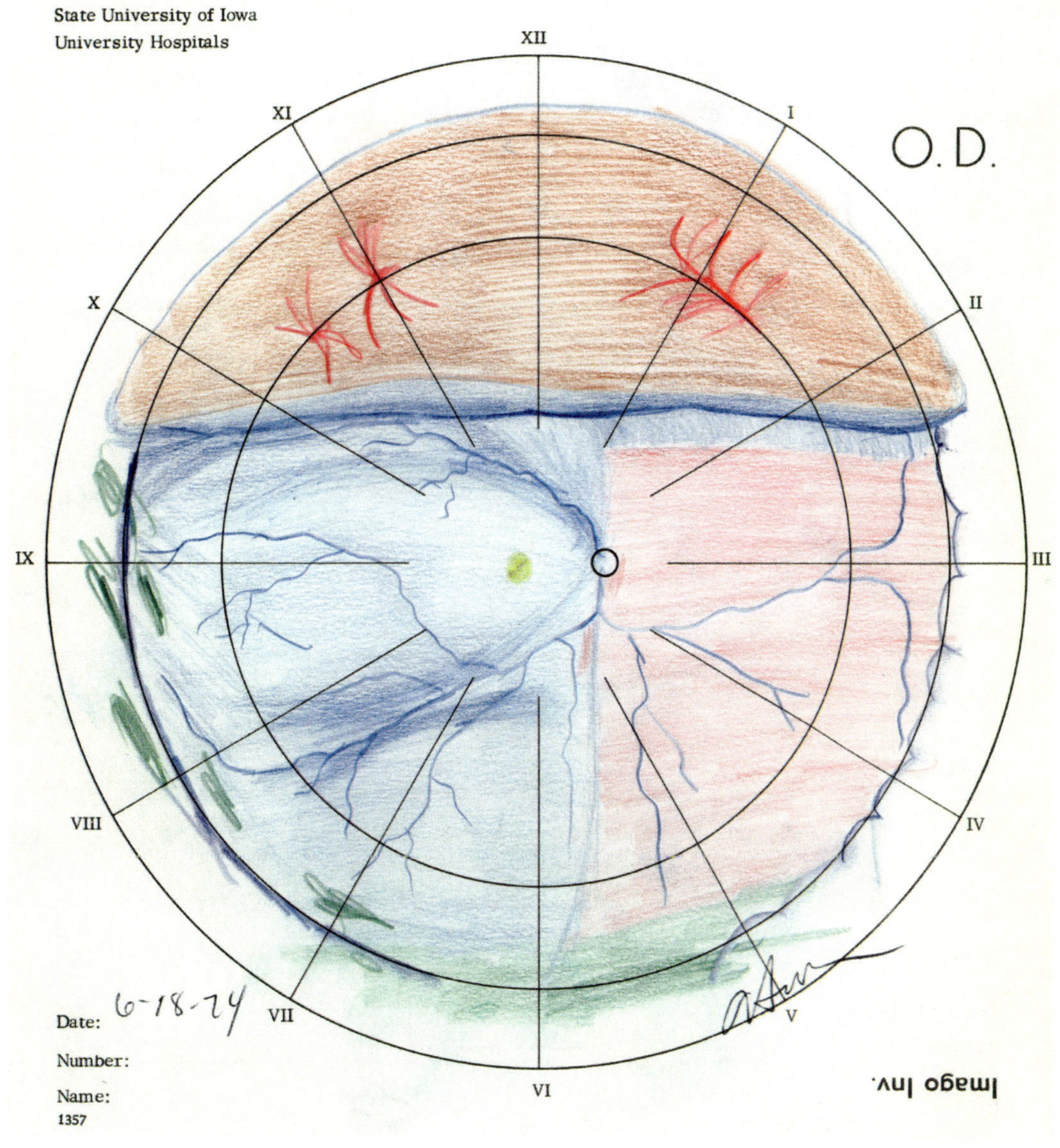

Artist Thomas S. Stevens
June 18, 1974

Diagnosis:
Giant tear of the retina, right eye, after being hit in the eye with oil pressure two weeks earlier.

"During those days [in a program with tiered rotations], the last six months resident became also a sort of junior staff, with a couple of elective rotations. Dr. Watzke didn't have a second staff member, and Dr. Tom Burton from U Illinois was just beginning his retina fellowship. Tom and I had a lot to learn, quickly, and Dr. Watzke probably aged ten years during that six months—but we sure got a lot of work and great experience."

David Townes, MD (12/7/10)

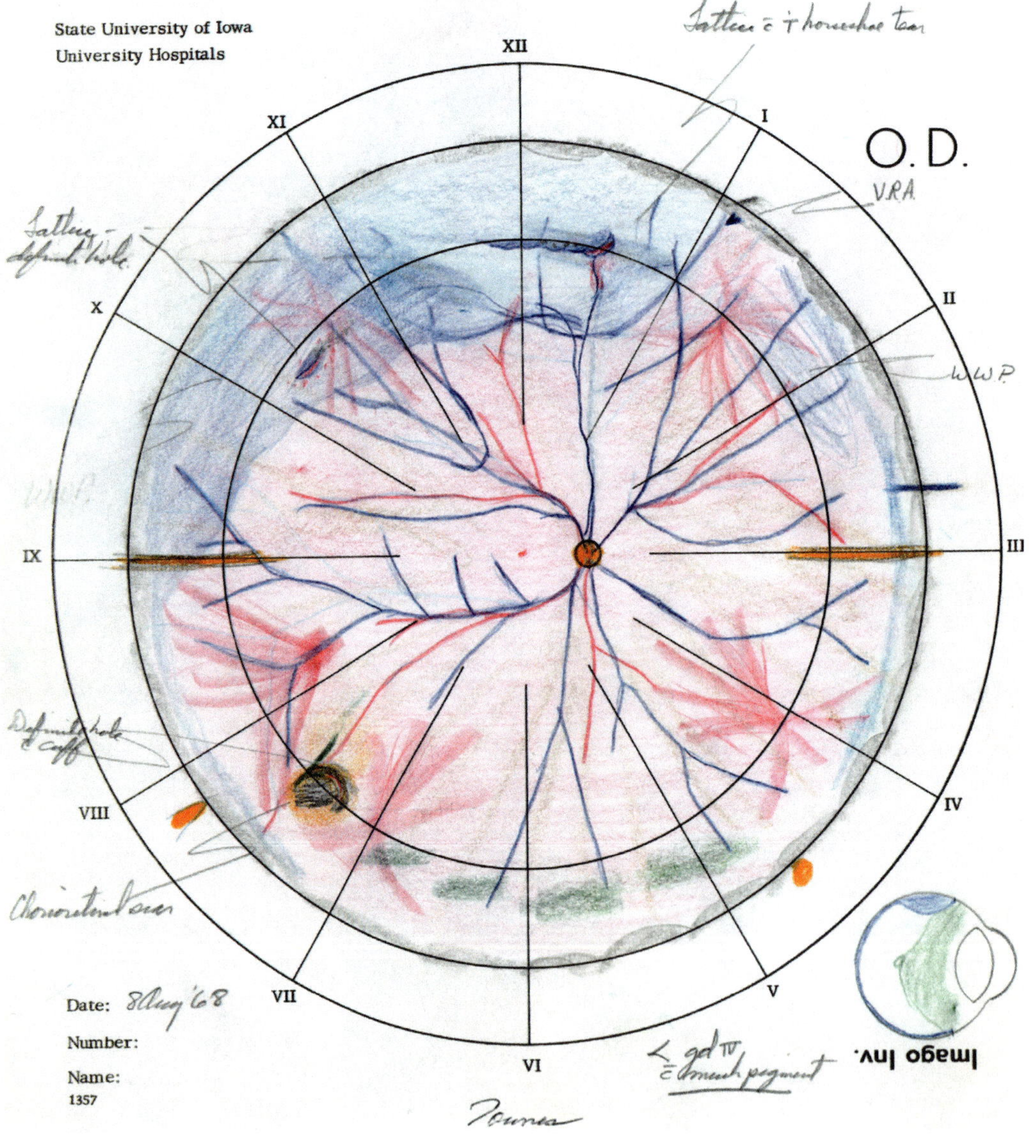

Artist David E. Townes
August 8, 1968

Diagnosis:
Retinal detachment, right eye, with two horseshoe tears noted in areas of lattice degeneration.

"I can attest to the fact that every *patient on the Retina Service had a retinal drawing made by the resident on the service."*

Melvin Rubin, MD (12/2/10)

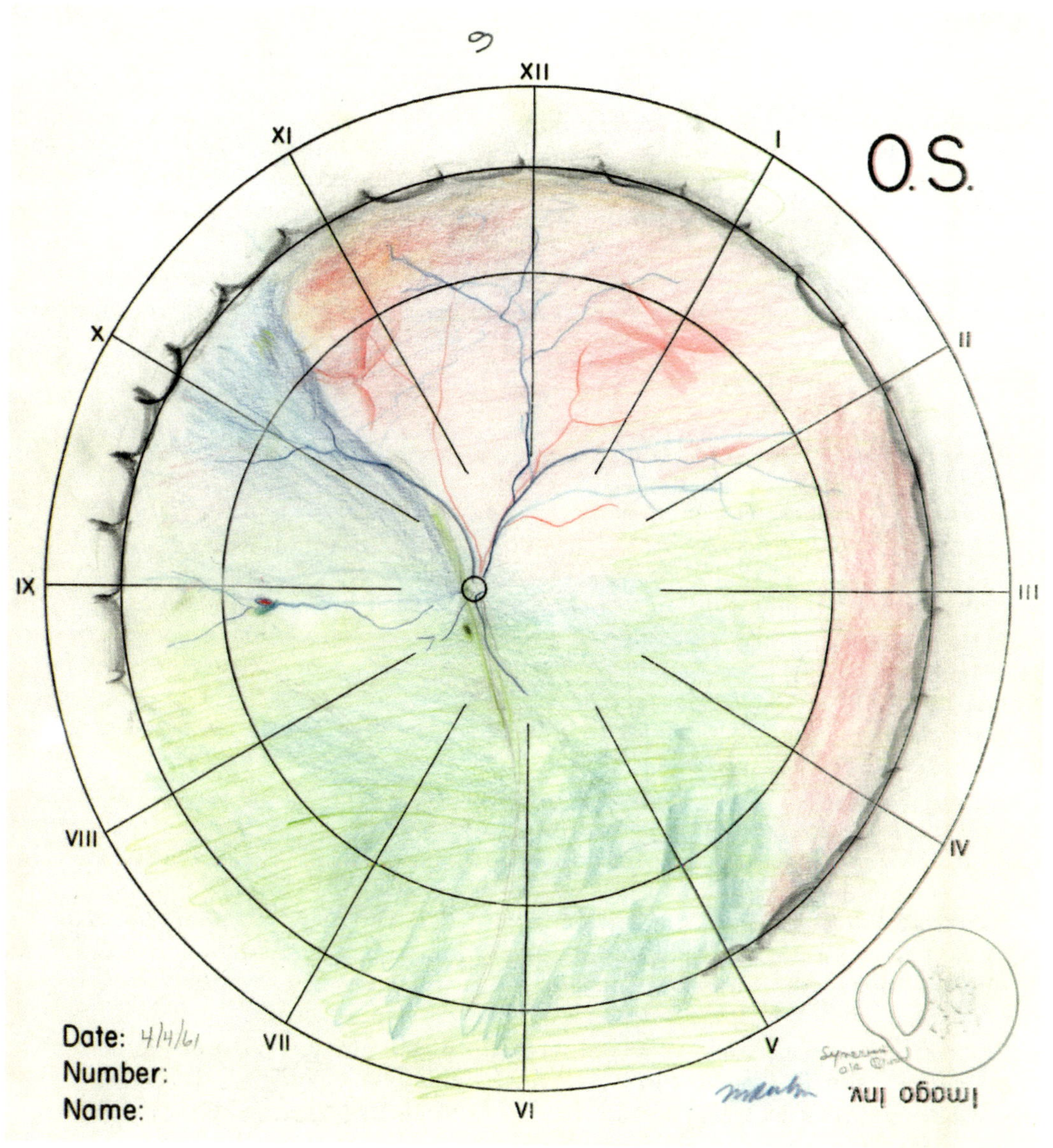

Artist Melvin L. Rubin
April 4, 1961

Diagnosis:
Retinal hole with a small retinal detachment, left eye.

Drawing Process

Retinal drawing was a skill of passage along each trainee's long path to mastery in understanding retinal diseases. For the residents, or the ophthalmologists in training, transitioning from general medical examinations using the direct ophthalmoscope to performing comprehensive ophthalmic examinations using an indirect ophthalmoscope required the skills needed to fully view the retina and plunged each new ophthalmology resident onto the intellectual and hierarchical bottom of the resident's newly chosen specialty. Ironically when student ophthalmologists begin life, they start out unable to see the retina. They learn each nuance and recognizable pattern gradually, however, if they persist long and carefully enough.

Why Examine "Indirectly"?

The indirect ophthalmoscope is so named because the image is seen "indirectly" through a condensing lens rather than being projected directly onto the examiner's retina. Multiple optical elements, which include the examiner's eye, a handheld lens, an illuminating head-mounted, binocular telescope, and the patient's eye or more specifically the retinal structure being observed, align to form the image. Because these elements work together in an upside-down and backwards (inverted) fashion when compared to the anatomy, the examiner must find a way to transform each additional field of view or vignette in order to fit it into the complete view as one would manipulate each puzzle piece to fit it into the whole. During an examination, the rotational distortion, combined with the spherical projection necessary to represent the curved surface of the retina on a flat drawing paper, introduces additional opportunities for perceptual errors and interpretation. Artistically, the method is most analogous to mirror anamorphosis, a method developed in the Renaissance in which people viewed paintings through cylindrical or distorting mirrors to challenge the skill of the artist and capture the interest of the viewer.

Most students of the medical art of indirect ophthalmoscopy needed a minimum of three to four months of daily practice to develop a working use of the indirect ophthalmoscope. The students needed even more skill to draw the retina, the structures under and around it (collectively referred to as the fundus), and the front or anterior periphery.

Because the anterior periphery lies in front of the equator of the eye, the examiner typically cannot visualize it with direct viewing and must use an additional technique called scleral indentation, in which the examiner puts pressure on the eyelid and indents the eye wall. The positioning of the

"I do remember that the retina division started training fellows when I was a senior resident. That meant that I had to give up some of my cases to the fellow. I remember one evening when I was assigned to work up a new retina case that meant staying late and doing the drawing. I didn't feel like it since I wasn't going to get to do the case and thought the fellow should do the drawing. The drawings were done mainly to force the surgeon to look at all parts of the retina and become familiar with the pre op anatomy. The fellow didn't think that was a great idea and let me know in no uncertain terms that it was my job."

John Ramsey, MD (12/6/10)

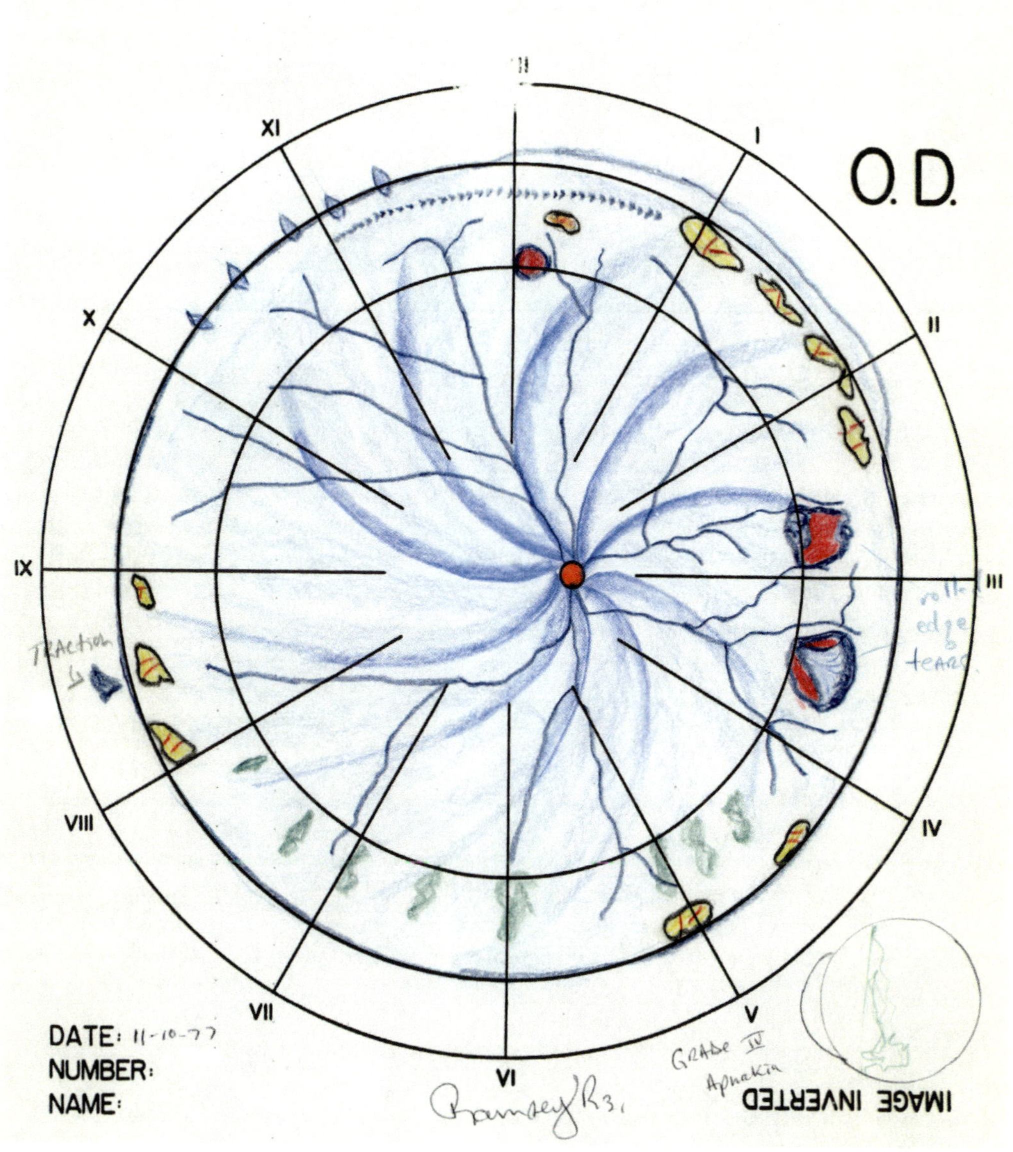

Artist John F. Ramsey
November 10, 1977

Diagnosis:
Retinal detachment, right eye, following cataract extraction.

indentation in alignment with the other optical elements and the retinal structure being observed requires additional physical dexterity. Again, because the method of drawing from indirect ophthalmoscopy was a complicated process, the examiner had multiple opportunities for creative and skill-related perceptual or perspective variation.

Performing an Indirect Retinal Examination

Just as a cardiologist would listen for heart sounds in a patient in a quiet room, an ophthalmologist would ideally perform examinations with an indirect ophthalmoscope in the dark. Especially for retinal drawing sessions, which could take a prolonged time, the patient would be comfortably lying in a face-up position at the height of the examiner's waist. Since examination was best accomplished from position 180 degrees from the retinal structure to be viewed, the examiner would position himself or herself around the head. Maximizing the examiner's retinal sensitivity facilitated viewing but would require up to twenty minutes in the dark to achieve sufficient dark adaptation. Following dilation of the pupil, the examiner systematically surveyed the retina to ensure that all of its area was inspected. Because of a patient's central retinal sensitivity, it was often easiest to examine the mid- and anterior periphery of the eye first and examine the optic disc and central retina last. However, as noted below, when the examination included formally drawing a retina, the examiner would draw the central retina first.

Drawing the Retina

Because of the optical inversion and projectional distortion of each retinal viewing, it was necessary to make some accommodation to integrate each field of view or vignette into a comprehensive retinal drawing. Most practitioners during the formal fundus drawing era used a mechanical approach in which the examining artist inverted the drawing paper and, starting with the optic disc and progressing outward, added each vignette by matching the retinal vessel pattern on one field edge, like matching the border of a puzzle piece.

The artist then progressively expanded the drawing from the optic disc to the periphery, one retinal image at a time, as one would do if constructing a puzzle starting from the center and adding pieces outward rather than following the usual frame-to-center approach. After drawing the

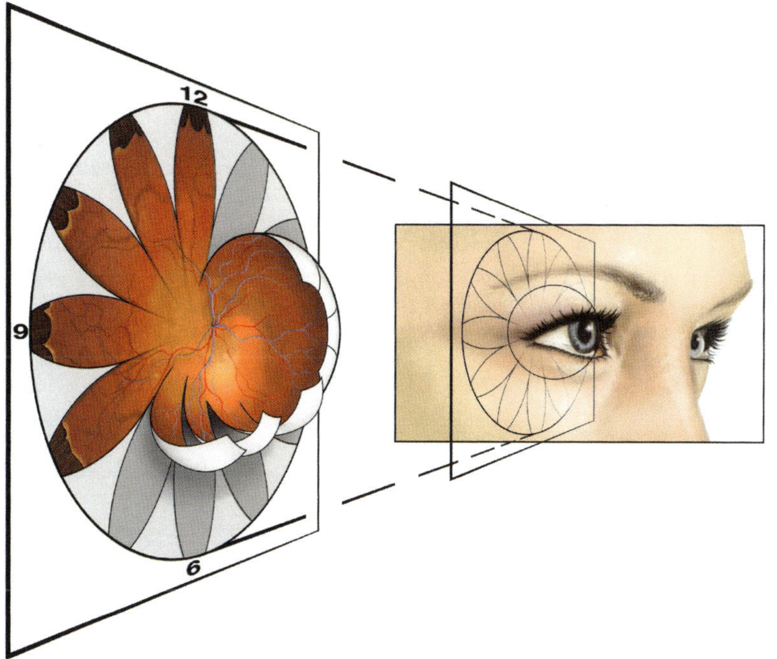

Figure 2. Illustration of the relationship of the retinal drawing to the patient.

vessel pattern throughout the fundus, the artist could check and confirm the orientation according to a clock hour by indenting the eye at twelve, three, six, and nine o'clock. As the artists developed a more experienced perception of this transformation or transposition of the entire retinal image, they could make simple drawings from the direct mapping onto an un-inverted paper, a skill similar to the mirror writing of Leonardo da Vinci.

While the artist was waiting for his or her own retina to adapt to the dark, he or she usually drew the major retinal vessels to the equator of the eye or fundus first and then added easily viewed landmarks and finally details of refined interest, such as retinal tears, pigmented spots, tumors, and major retinal folds.

For the final sketching stage, the artist performed scleral indentation, which was usually done with sufficient gentleness, given the goal was to avoid patient discomfort. Artists followed a systematic approach for scleral depression to inspect all regions of the peripheral retina and locate any retinal breaks or other clinically relevant abnormalities. Typically an artist began depressing the eyelid (and globe) at twelve o'clock, and progressed laterally to six o'clock and back to twelve o'clock, moving either clockwise or counterclockwise, depending on whether the right or left retina was being drawn. Landmarks that may have easily been seen, even through cloudy media (the tissues and cavities through which light passes), were especially important to add at this stage.

Refinements of the Drawing

Once the artist completed a sketch or preliminary drawing, he or she could enhance it with additional shading and without the need for the patient as model. Depending on the interest, commitment, and attention to detail of the individual artist, or the rules set by that person's supervising faculty member, the artist could add variations in line, shading, color, and other components of each drawing. Elective embellishment might add another hour but contributed richly to the artistic process and appeal.

Teaching Benefits and Requirements for Retinal Drawing

One of the more challenging aspects of retinal drawing was in teaching, guiding, and mentoring the examining artists. Already within eye care, management of diseases of the retina and fundus was considered a particularly difficult task. Even seasoned professionals encounter frustration and difficulty when trying to view the tissues necessary to diagnose and monitor management of the retina. Indirect ophthalmoscopy required precision and persistence, and yet the time needed to develop indirect ophthalmoscopic examination skills often tested the limits of patience of examiners and their instructors. Adding months of additional time to develop a working skill in retinal drawing often exceeded the enthusiasm of residents. However, once the examining artists could convince themselves that they could piece together fields of view and see the alignment of retinal vessels to the equator, a renewed sense of pride in their new skill usually increased their enthusiasm.

In institutions where retinal drawing was an expectation, teaching others the skill of retinal examination became more gratifying when the examiner-in-training succeeded. The process of committing to a diagnosis for each feature and its location created the opportunity for learning or confidence building, which could be affirmed by the attending faculty. As artists compared their interpretations of retinal variations and abnormalities to the interpretations of faculty, or at times to those of a fellow or senior resident, all present examiners could have their decisions confirmed or refined. Each examiner could evaluate whether he or she could recognize each subtle abnormality as well as the senior examiner. Teaching had become collaboration.

"Retina clinic was in the 'bullpen,' a large room with three stretchers on each side. Multiple simultaneous exams created quite a hubbub of activity. Lights dimmed, multiple indirect ophthalmoscope beams created a somewhat surreal setting—like a laser scene from Star Wars. The retina fellows were fast examiners and drawers, the residents, not so fast. But detail was critical and we all wanted approval of our work. I remember Dick Dreyer critiquing one of my first drawings. 'Pretty good,' he said, 'but this tear is a bit over this way,' as he penciled in a horseshoe tear about a quarter inch left and an eighth inch more peripheral than my depiction. That's when I realized I'd better be really astute! Thus the hour-long sessions…"

Francisco Pabalan, MD (12/6/10)

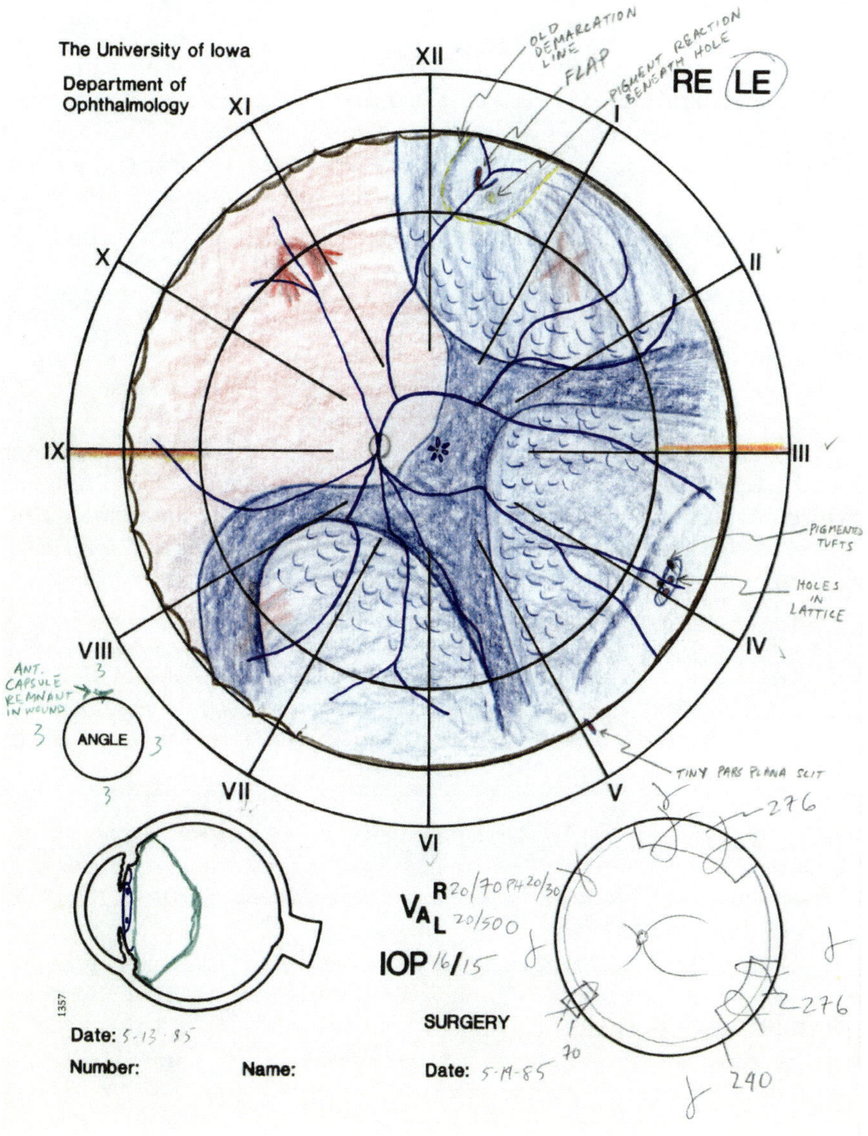

Artist Francisco J. Pabalan
May 13, 1985

Diagnosis:
Bullous retinal detachment and a horseshoe tear, left eye, seven months following cataract extraction.

Influences

When Dr. Jonathan D. Fratkin replied to a question about his authorship of a December 1976 image, he included an explanation that concerned collaboration, specifically how he had started the drawing in question, but probably his superiors added to or otherwise revised it. This situation seemed different from faculty pointing to a tear that a resident had missed. It also seemed different from the point in the history of UIHC fundus drawing when the lengthy examinations and documentation happened in a retina room and various physicians saw the retinas and pathology of colleagues' patients. In neither of these usual circumstances was authorship in question.

It had been clear from months of research that drawings done by a particular artist followed that artist's style and could be identified according to the artist's name—for example, "This one is a Diamond" or "Should we include another Vermillion?" This familiarity of style, especially when an artist had several drawings in the bunch, was noteworthy but not surprising. Writers and musicians also have their own style.

When certain drawings looked so much like others, and it was found that the artists were actually *not* the same person, we considered not only collaboration but also influence. Outside influences, like art classes, and internal influences, like talent, affected the drawings; however, the question of who influenced whom when the artists were classmates surfaced in a few instances when the artists' styles were so similar that, at first glance, the viewer would assume only one artist was responsible for the two images.

Dr. H. David Fenske's and Dr. C. Neal Jepson's images from August 1963 and May 1962 reflect this phenomenon. Details, shading, and even the quality of annotations show similarities of style, although different artists did the drawings—a fact checked through review of medical records and confirmation by the artists. Dr. Fenske's remark about trying to outdo each other (see 8/22/63 image) attests to the fun competition that came with collegiality.

Dr. Patrick J. Caskey (see November 1986 image), in his comment about the similarities between his drawings and those of Dr. Patrick Coonan (see August 1987 image), suggested collaboration between classmates and influence from the works of previous fellows. However, Dr. Richard F. Dreyer said influence from predecessors was not always the case, as "the old drawings were not necessarily available to us" (3/1/11).

Why then, the question could be and has been asked, do many of the drawings reveal an influence that is less likely from imitation and more likely a sign of the times? In the following chapter, the similarities of certain drawings expand beyond two residents or fellows whose time on the Retina Service at UIHC overlapped. Sometimes styles, albeit not as strikingly similar as the preceding pairs, cover an entire decade.

"As a newbie in the Retina Clinic in '76, I provided a Chevy chassis that someone else (perhaps the fellow, more likely staff [e.g., T. Weingeist] used to build a Mercedes). The 3-D effect, probably achieved by wetting the blue pencil (?), would have been too advanced for me. However, I can recognize my handwriting: the name and the date. You certainly can have all rights and privileges to my portion. Indeed, since I own the foundation, I technically own what was built on my property."

Jonathan D. Fratkin, MD (12/6/10)

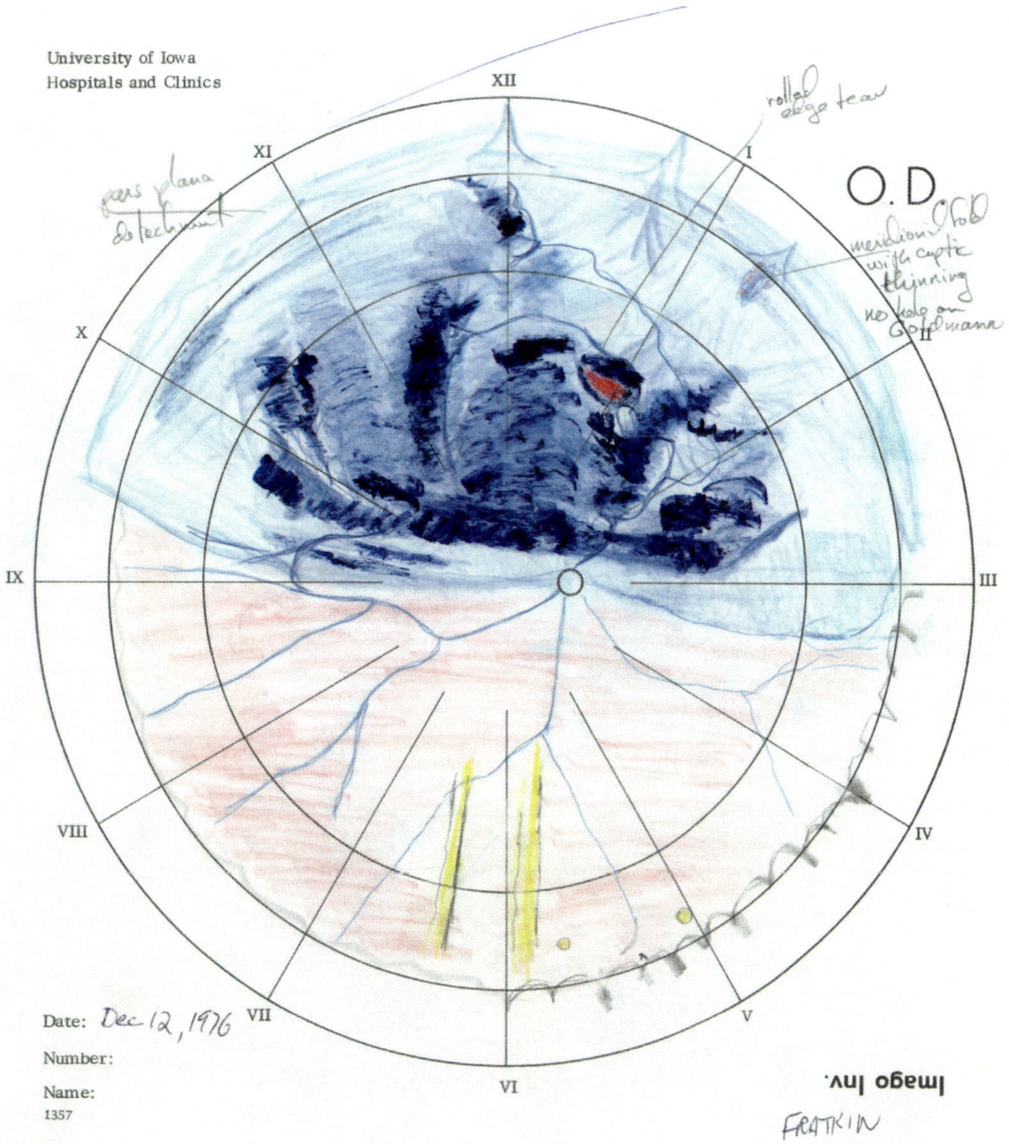

Artist Jonathan D. Fratkin
December 12, 1976

Diagnosis:
Bullous detachment with large flap tear, right eye.

"Classmate Neal Jepson and I spent much much time—poor patient—making these attempted accurate drawings on patients, trying to be as accurate as possible, considering our somewhat early knowledge and first exposure to this subspecialty in our training. We took great personal pride in trying to outdo each other in accuracy and artistry."

David Fenske, MD (12/6/10)

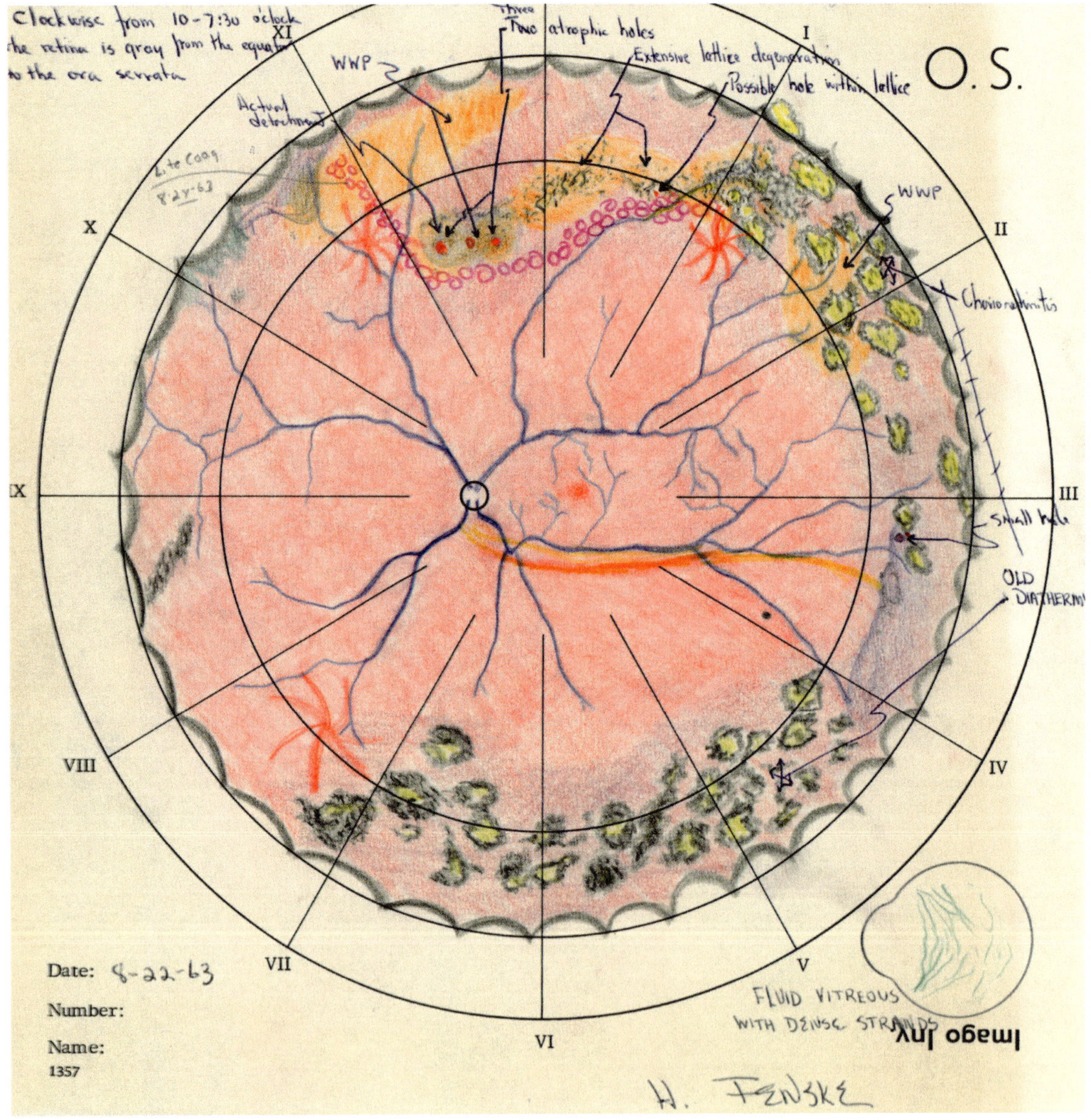

Artist David Fenske
August 22, 1963

Diagnosis:
Lattice degeneration and atrophic hole with flat detachment, left eye.

"Neal did influence me a lot in creating these colored renditions of the retinas, and residents who followed us tended to do the same thing many times."

David Fenske, MD (12/6/10)

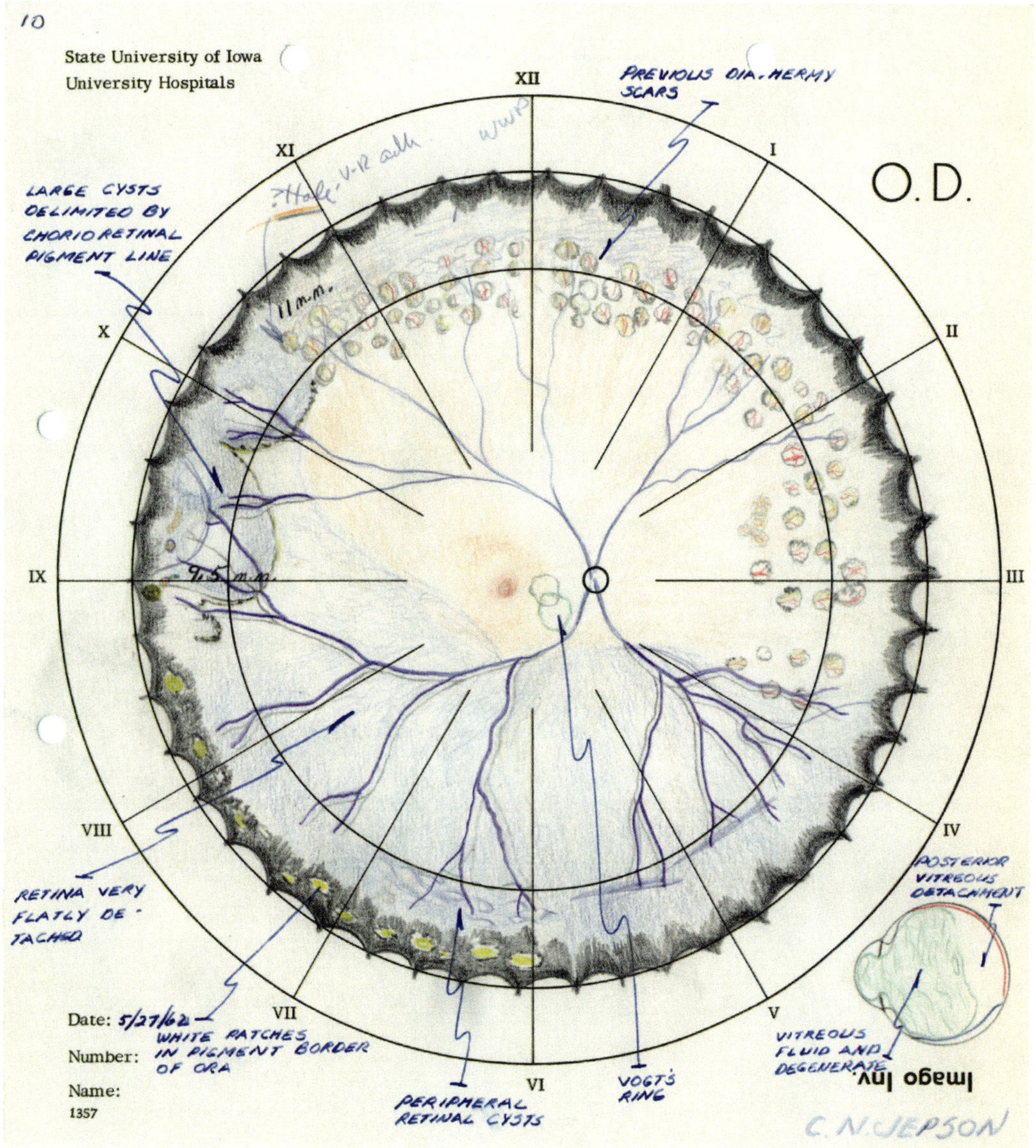

Artist C. Neal Jepson
May 27, 1962

Diagnosis:
Recurrent retinal detachment, right eye; cataract extraction seven years earlier.

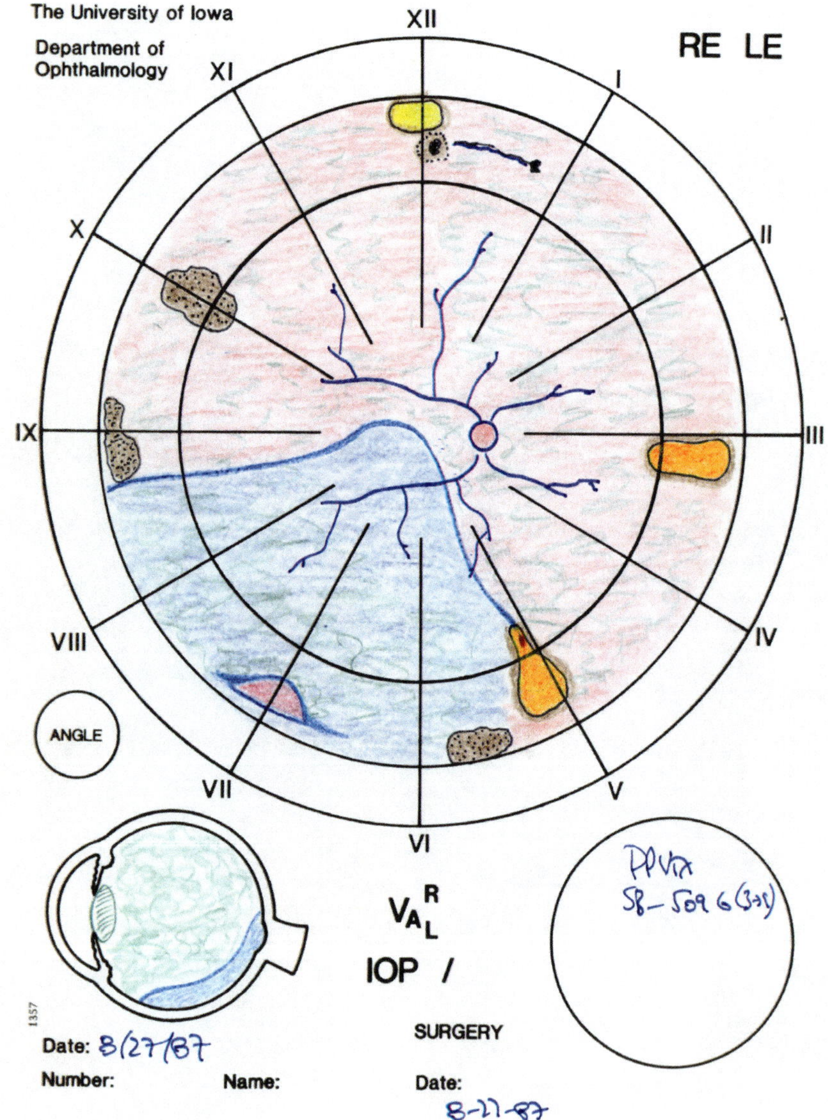

Artist Patrick Coonan
August 27, 1987

Diagnosis:
Retinal detachment, right eye, probably secondary to traction from chronic inflammation, most likely toxoplasmosis.

"Patrick Coonan was the fellow that arrived at Iowa six months after I started and wound up being my partner for the first eighteen years of my practice! It would make sense that his drawings look like mine since we all looked at the drawings of the fellows before us to get an idea how to go about creating our own."

Patrick Caskey, MD (12/6/10)

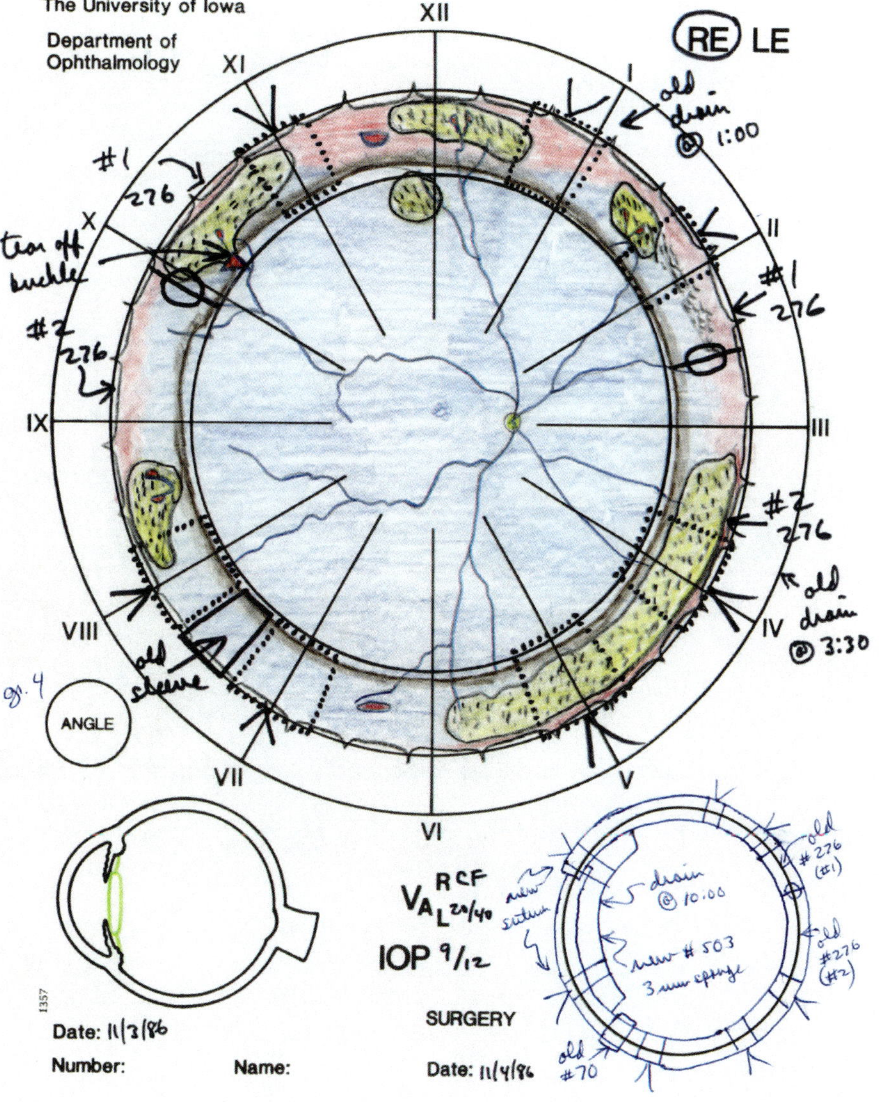

Artist Patrick J. Caskey
November 3, 1986

Diagnosis:
Recurrent retinal detachment, right eye.

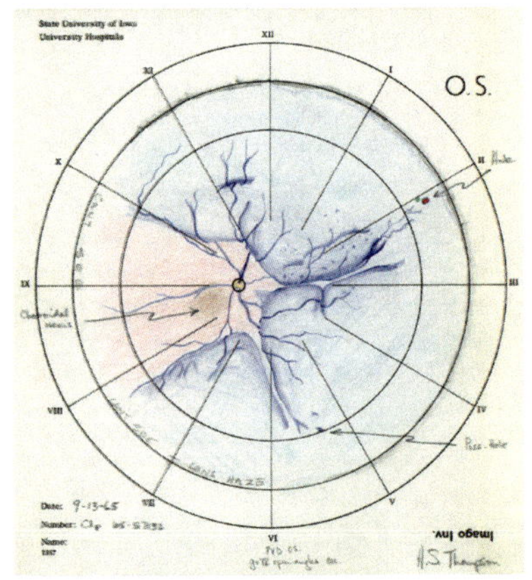

1960s
Realistic Embellishment

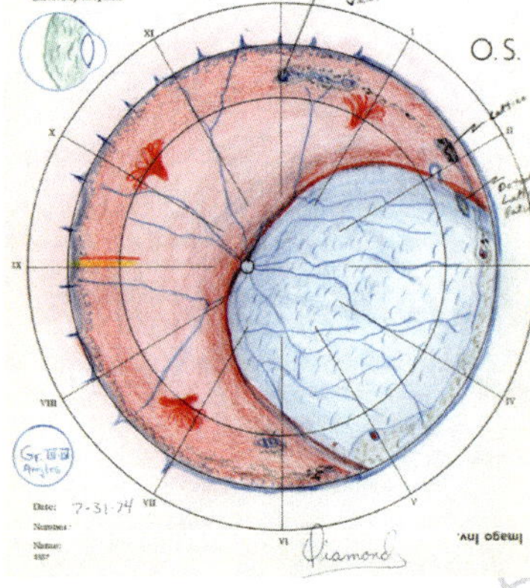

1970s
Modern Expression

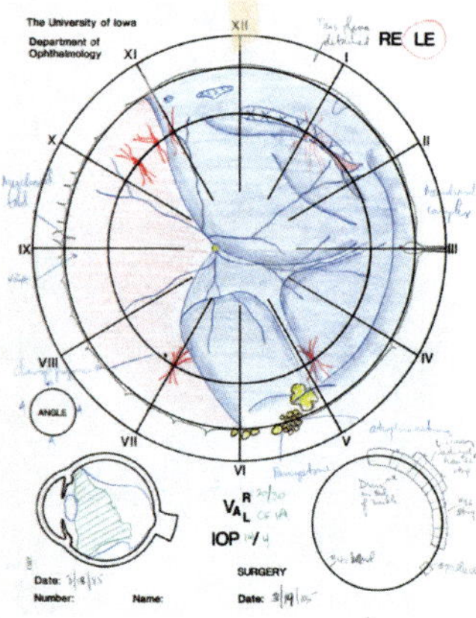

1980s
Iconic to Symbolic Caricature

Chapter 2

How Group Drawing Styles Changed over Time: Shared Artistic Interpretation

One of the most interesting and subtle observations made during the examination of the images' stylistic elements was that, over the course of more than thirty years of fundus drawing, the drawings evolved in their general appearance and style.

Although each individual's personal interpretation differed, most images created within a given period share enough similarities to be recognizable by membership to a group style. Further, these groupings of drawings fall into easily distinguishable categories, which seemed to change over roughly ten-year intervals.

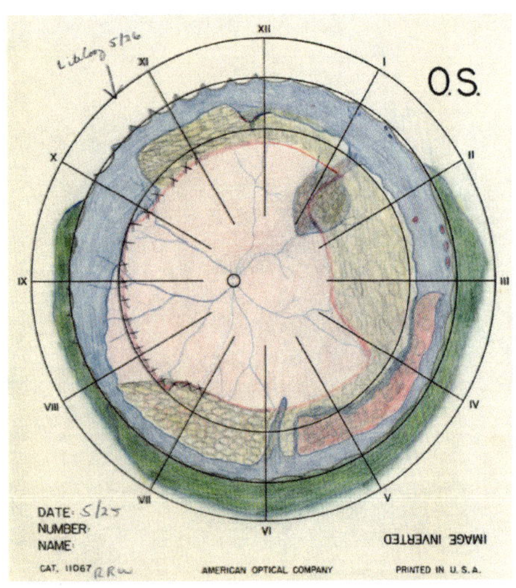

Russell R. Widner, 1960s
(See page 50 for full-size image)

It is striking that each of these themes followed a progression from more complex to simple or iconic over time. For example, early drawings from the 1960s appear more realistic, depicting numerous normal and pathological fundus features in location, relationship, and proportion, but not in the actual colors that the ophthalmologist saw, due to the use of a color code. A decade later, in the 1970s, the drawings resembled modern art with larger forms, bolder colors, and simplified details. By the 1980s, the drawings became more stylistically symbolic, whimsical, imaginative, and at times minimalist in nature.

Although there may be multiple reasons for why the drawing style evolved, or devolved, the way the drawings changed adds to one's appreciation of the stylistic variation and inclusion of artistic interpretation. For example, included here are two images, one from Dr. Russell R. Widner (left, from the 1960s) and one from Dr. Samuel G. Farmer (right, from the 1980s) that epitomize the range in representation. Depending on a reader's perspective and sense of humor when comparing these images, he or she may feel the changes are for better or worse, given the artistic or documentary purpose.

Advancements in medicine led the artists to focus their drawing on pathology rather than on the entire fundus. Why the stylistic elements in the drawings changed, however, is less apparent. Group shifts in representation may have been due to particularly influential individuals or changes in illustration emphases by faculty. Alternatively or synergistically, changes in patient management, documentation needs, the time available for drawings, or the sensibility of the author and patient may have contributed.

In this chapter, each epoch is featured within a separate section to highlight common aspects, and each section is titled based on its most recognizable attributes. Although no direct comparisons were made to fine-art stylistic evolution, to some degree the changes among retinal drawings seemed to parallel stylistic evolution in the wider artistic context.

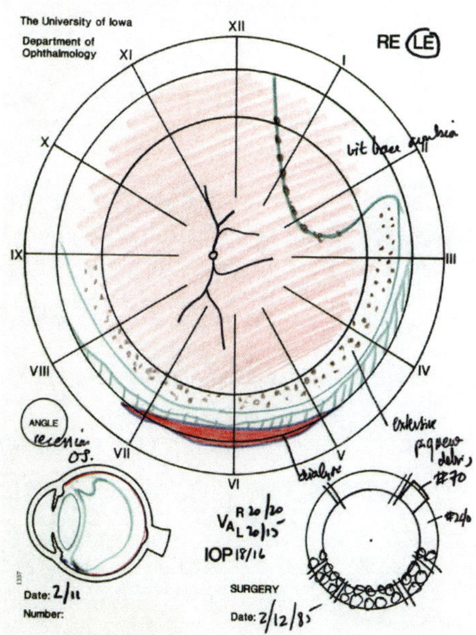

Samuel G. Farmer, 1980s
(See page 79 for full-size image)

Breton's scenes of country girls "engendered a 'naturalism completely infused with poetry,' as [French naturalist writer Emile] Zola defined it."

Laura Lombardi, art scholar

Jules Breton

The End of the Working Day
(Fun du Travail)
1887, Oil on canvas
New York City, Brooklyn
Museum

Realistic Embellishment

Retinal drawings done during the late 1950s and 1960s, which have a realistic or embellished style, recall the art schools of Naturalism and Realism. These schools arose in nineteenth-century Europe. Naturalist and Realist artists endeavored to depict what they saw. Their works replicate the natural or real qualities of the scene, as it was observed. Works created during this era are at times beautiful versions of what the artist saw, and at other times gritty and ugly, albeit amazing depictions, with variations on each. In these two movements, and in works of naturalistic art since then, artists aim to represent, with a degree of honesty and truth, objects and scenes seen every day. And so it is with many of the fundus drawings done throughout the 1960s.

Dr. Jepson, who referred to his works during that time as "my old ('pre-historic') retina drawings," recalled his method and the reasons for it: "I always could make accurate representations of what I saw, and since most of the other residents admittedly couldn't draw a straight line and literally hated the part of the Retina Service which required them to draw live retinas, I decided to spruce my drawings up with a touch of reality" (11/15/10).

Dr. Jepson's ability to accurately represent on a fundus chart the anatomy visualized with the help of an indirect ophthalmoscope, and the time and talent involved in "sprucing up" the drawings "with a touch of reality," led to his renown at the University of Iowa. His contemporary Dr. Fenske declared, "Neal was and still is the champion" (12/6/10). Many other physicians who studied at Iowa agreed, mentioning Dr. Jepson—either by name or by reputation—in correspondences connected to this volume. Dr. Baller, for example, stated, "As a senior resident, C. Neal Jepson was a perfect example, combining artistic nuances to critical retinal drawings. I tried (never succeeded) to make my drawings better than his" (3/25/11). Dr. Burton even added a somewhat critical spin to his comment, saying that, yes, Dr. Jepson did "outstanding" drawings, but they were "so elaborate that it was hard to get any information down from them" (12/14/10).

The realistic style neither began nor ended with Dr. Jepson. Realistic fundus drawings done by Dr. Watzke and Dr. Richard O. Schultz in 1960 actually predate Dr. Jepson's work. In addition, Dr. Fenske, who was Dr. Jepson's friend and classmate, as well as Dr. Widner, Dr. Baller, Dr. Otto A. Wiegmann, Dr. Bruce E. Spivey, Dr. H. Stanley Thompson, Dr. James M. Hersey, and Dr. Frank H. Reuling, produced drawings that realistically reflect the beauty of the retina as it was seen during examinations throughout the 1960s. Some used authentic coloring; some more closely followed the color code for the fundus chart. But these artists chose to draw what they saw, with minimal symbolism.

Realistic art has figurative or representational aspects. Although it is relatively more objective, in that the figures drawn resemble figures observed with less obviously interjected interpretation by the artist, the works carry a message: human life, even with suffering and strife, is visually captivating. The details and muted tones, whether muddy and earthy or moderately lit, convey that message, which was especially clear in the 1960s realistic movement of the fundus drawing era.

For the first time, ophthalmologists at Iowa were able to visualize the peripheral fundus to treat retinal detachment. Scleral buckling had revolutionized treatment for retinal detachments, but the procedure had been expanded to include a number of variations in technique and a host of implants to produce the needed globe indentation for retinal holes or breaks in various locations and arrangements. The more detailed and more accurate the drawing, the better the surgeons could plan the procedure and evaluate their implant or technique options. Differences in success were often attributed to millimeter differences in scleral buckle placement relative to the retinal holes. Such precision encouraged more detailed and precise drawings.

"You have to remember, at that time the binocular indirect ophthalmoscope was new to ophthalmology and was more like a middle ages torture skull-ring clamp," explained Dr. Robert J. Thompson (5/25/11). Because the peripheral retina could be seen with the modernized indirect ophthalmoscope ("the indirect" for short), tears in that area could be repaired and certain reattachments, which were previously impossible when the tears were out of sight, could be done with great success.

The fundus drawings that accompanied the indirect ophthalmoscope and were done preoperatively "had to be of one language if they were to mean anything," Dr. Thompson said. "Everyone had to draw the retina findings according to the rules/definitions to be able to be understood universally and historically."

Dr. Thompson went on to say, "Dr. Watzke possibly brought those drawing rules from his training with Dr. Schepens, and may have even developed them himself. He is the only one I know that has emphasized their detail and accuracy." Although the "rules," including the colors used, were necessary for documentation of pathology and communication concerning it, Dr. Thompson aptly noted that fundus drawing was "in constant change from the beginning" (5/26/11).

"I was enrolled in the Iowa art school before I was drafted into the Korean War, and after the war and my discharge, I decided to try premed with my $135 per month GI bill."

C. Neal Jepson, MD (11/15/10)

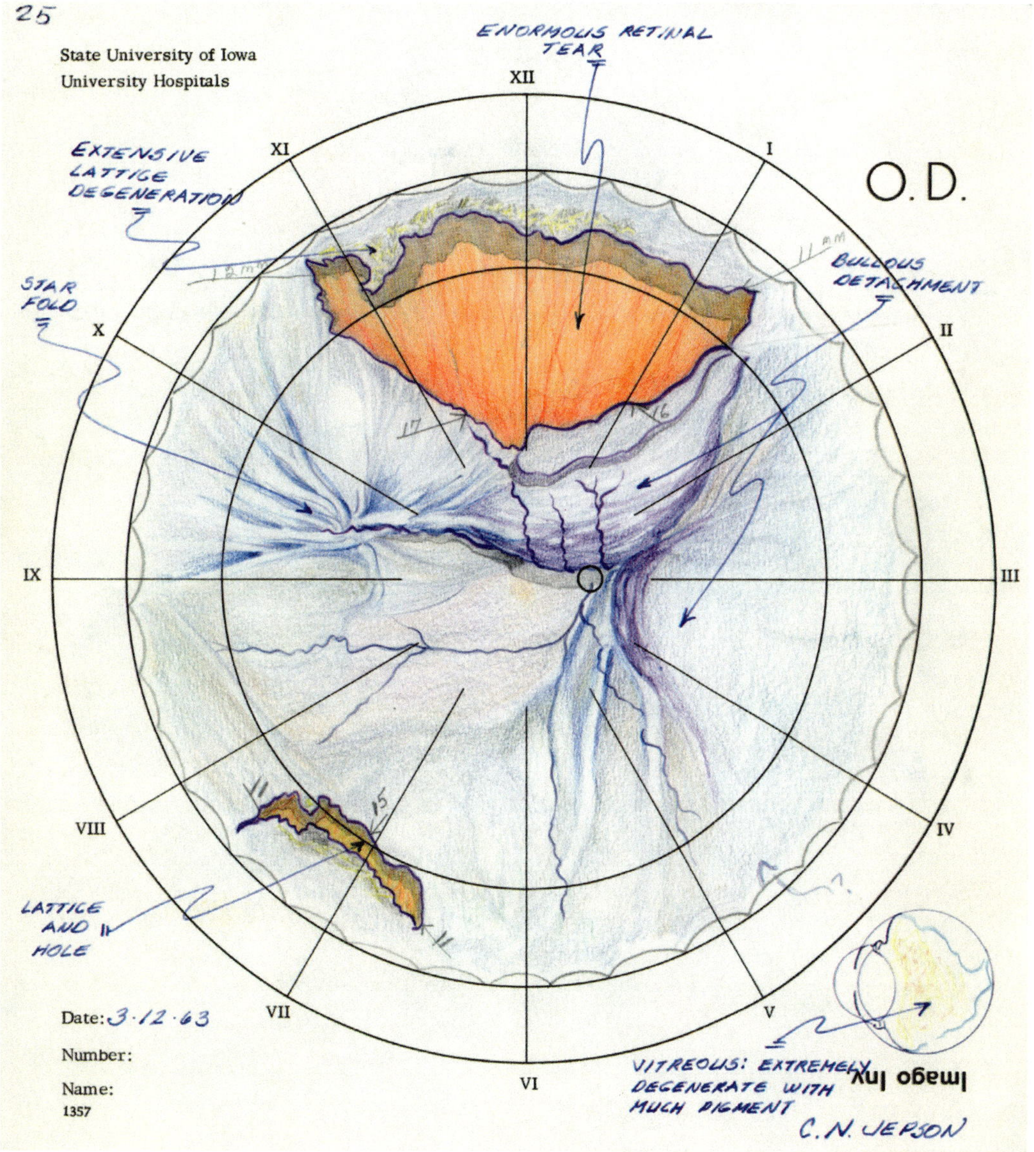

Artist C. Neal Jepson
March 12, 1963

Diagnosis:
Bullous retinal detachment and "enormous" tear, right eye.

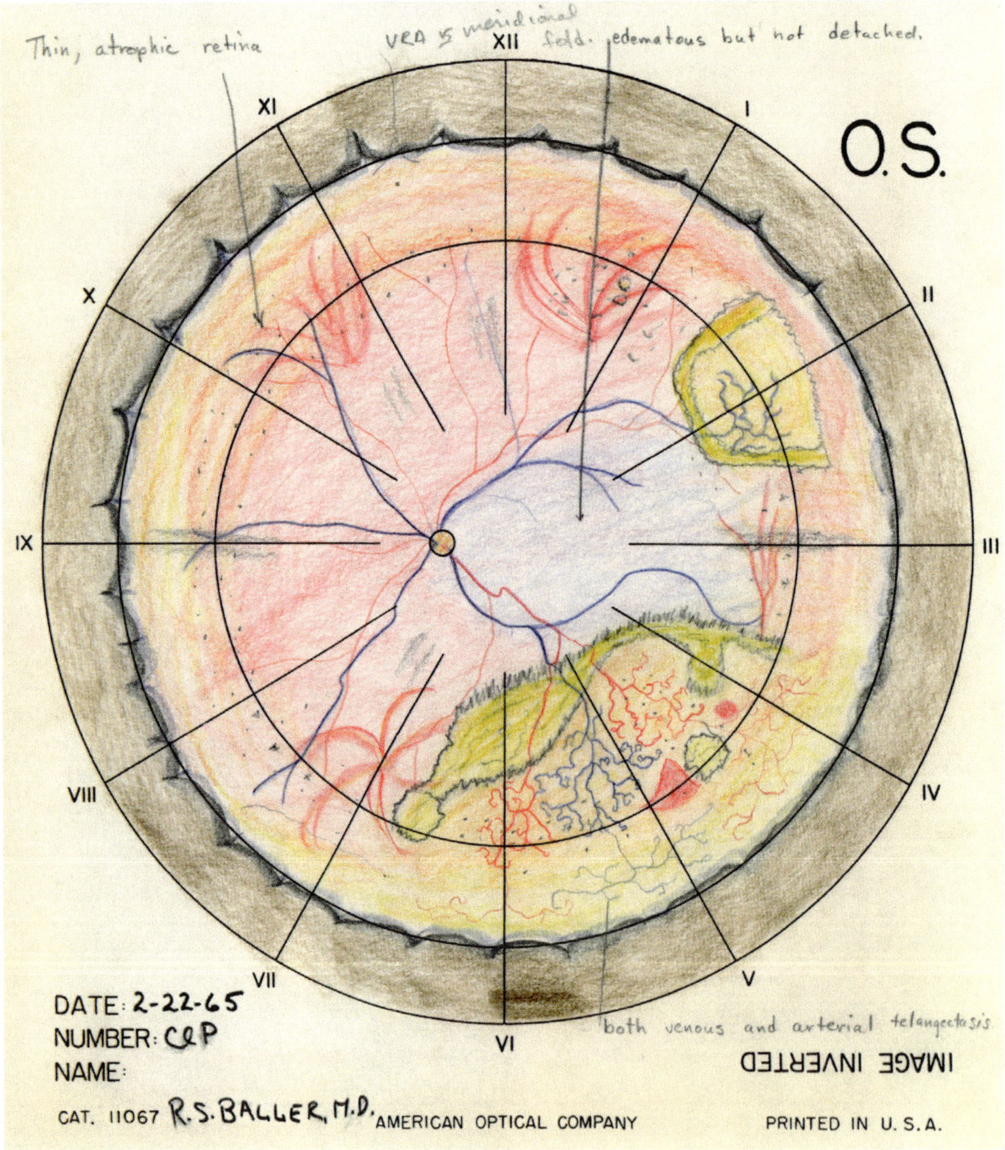

Artist Robert S. Baller
February 22, 1965

Diagnosis:
Coats' disease, both eyes, with left eye pictured.

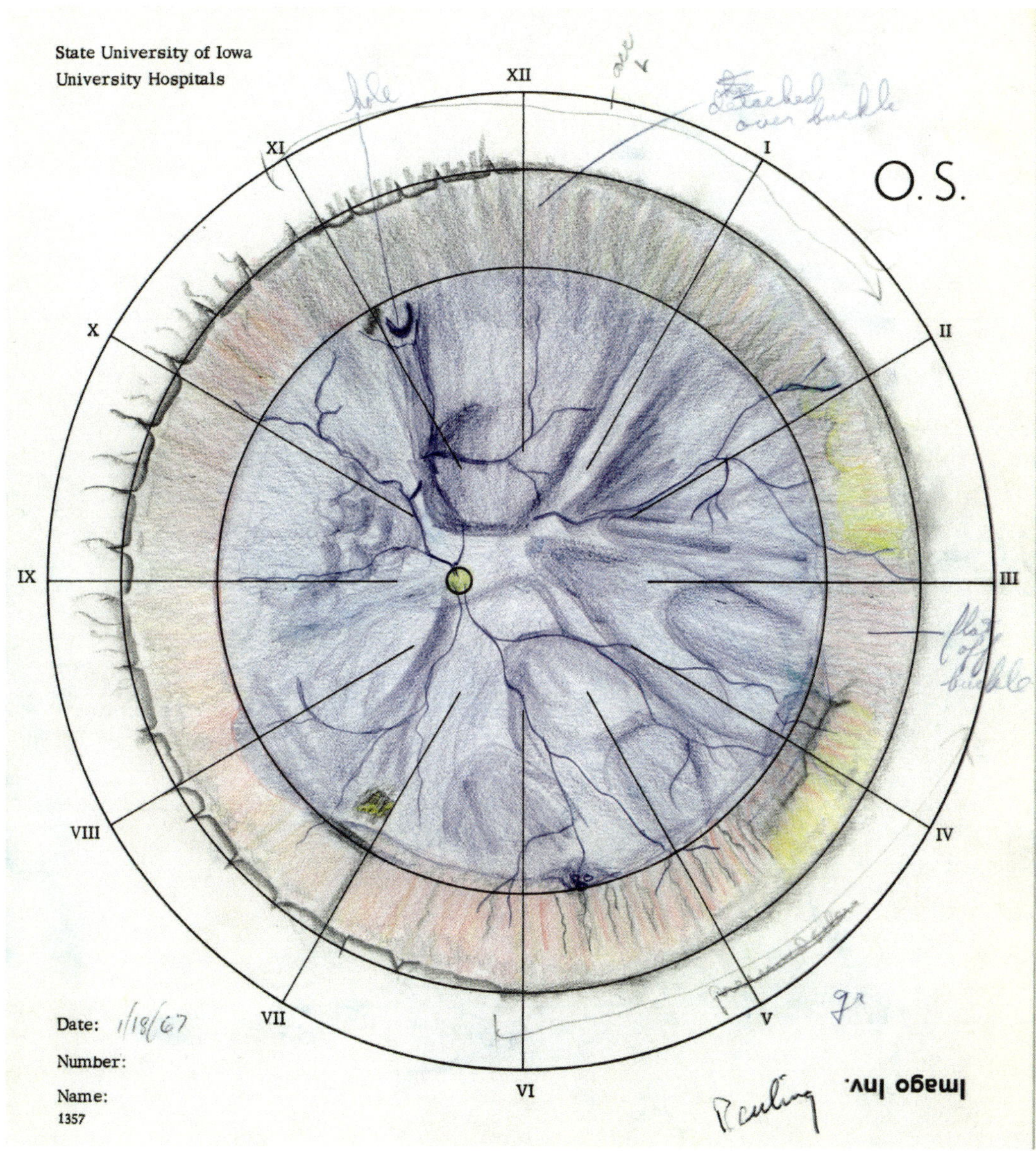

Artist Frank H. Reuling
January 18, 1967

Diagnosis:
Recurrent retinal detachment, left eye: "Retina off. Thrown into multiple folds."

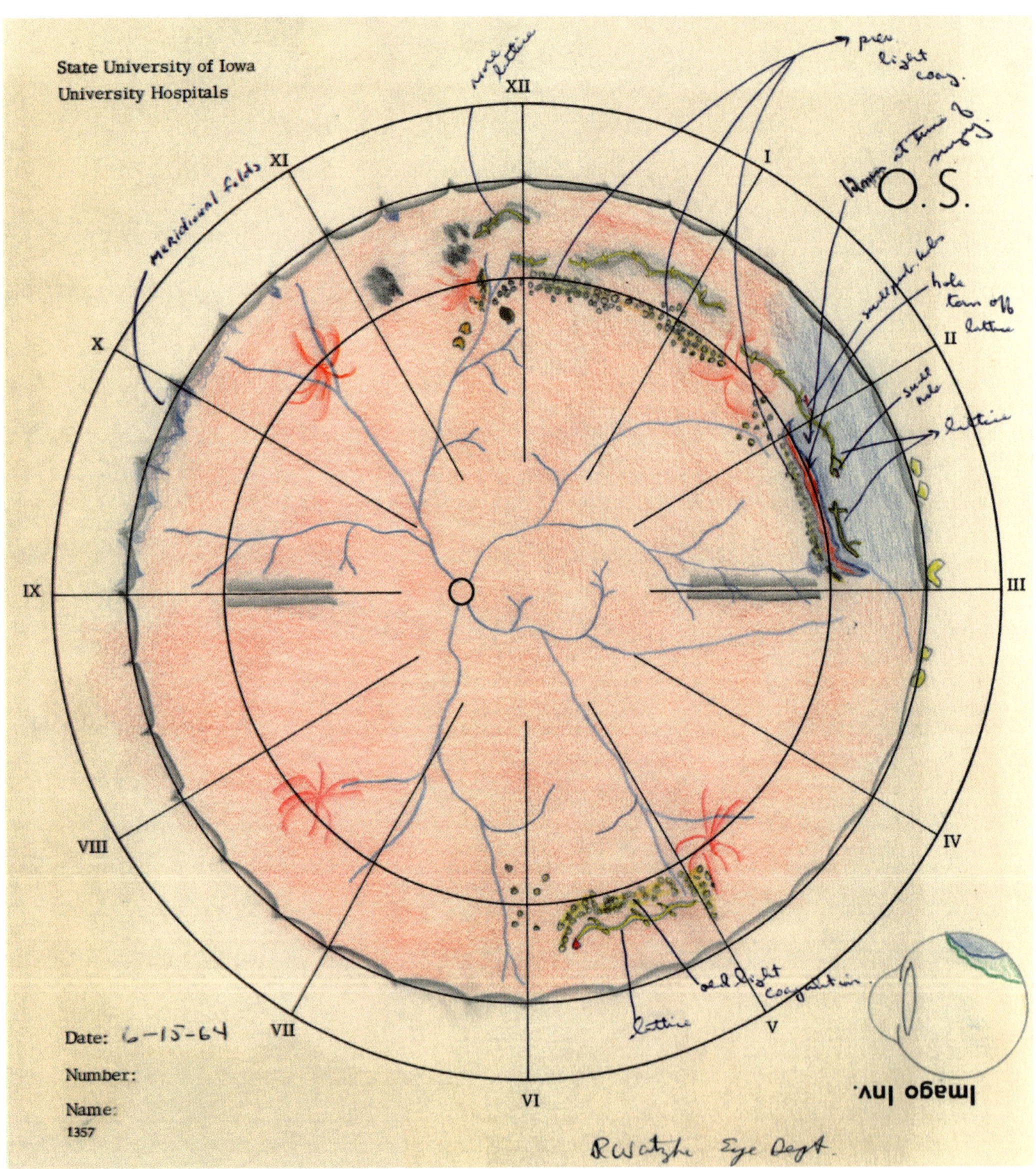

Artist James M. Hersey
June 15, 1964

Diagnosis:
Retinal detachment secondary to lattice degeneration, left eye.

"[A drawing] helps to explain to the patient what is going on and what therapy program is being planned"

David Fenske, MD (12/6/10)

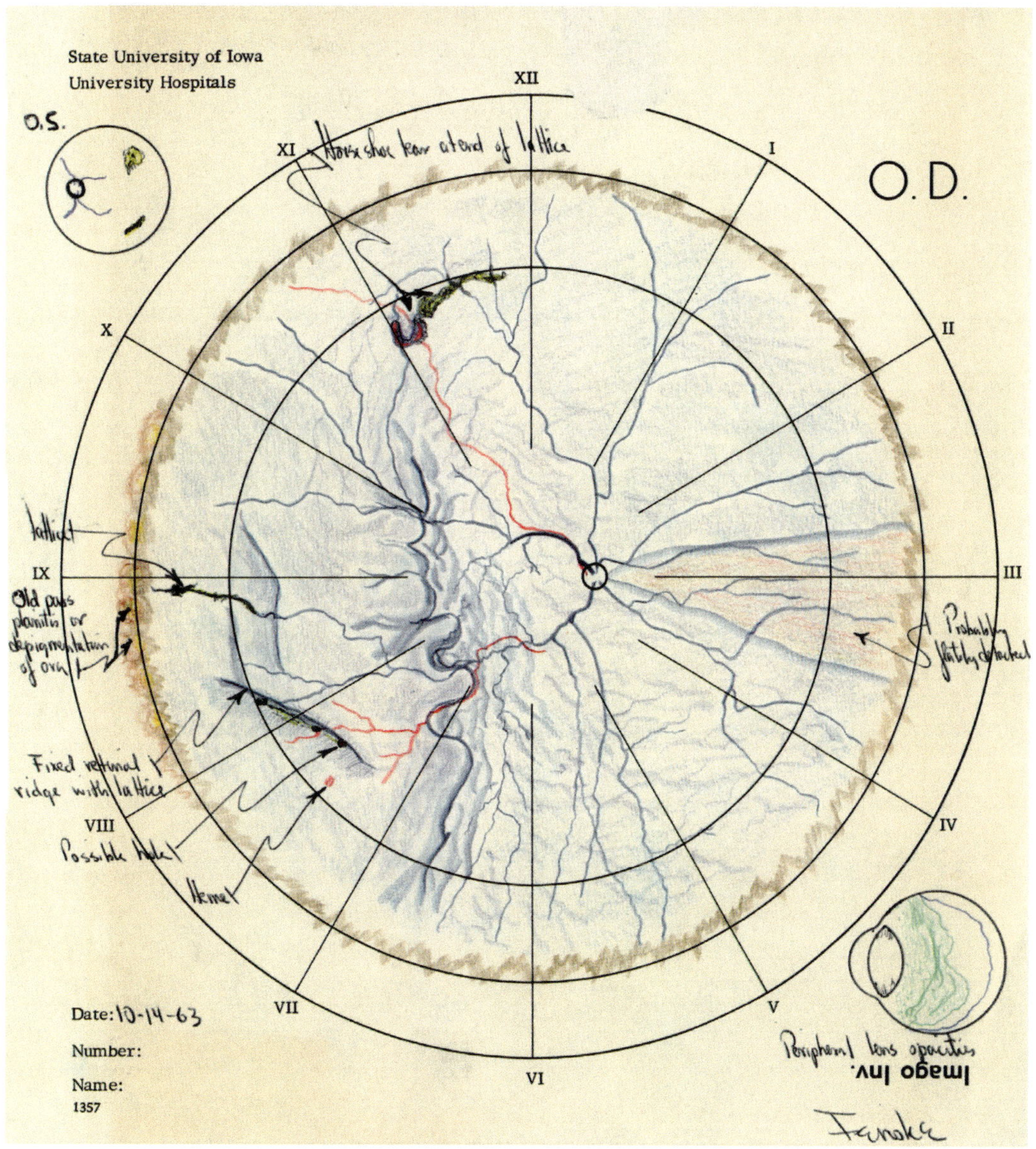

Artist David Fenske
October 14, 1963

Diagnosis:
Retinal detachment, lattice degeneration, large horseshoe tear, and small atrophic hole, right eye.

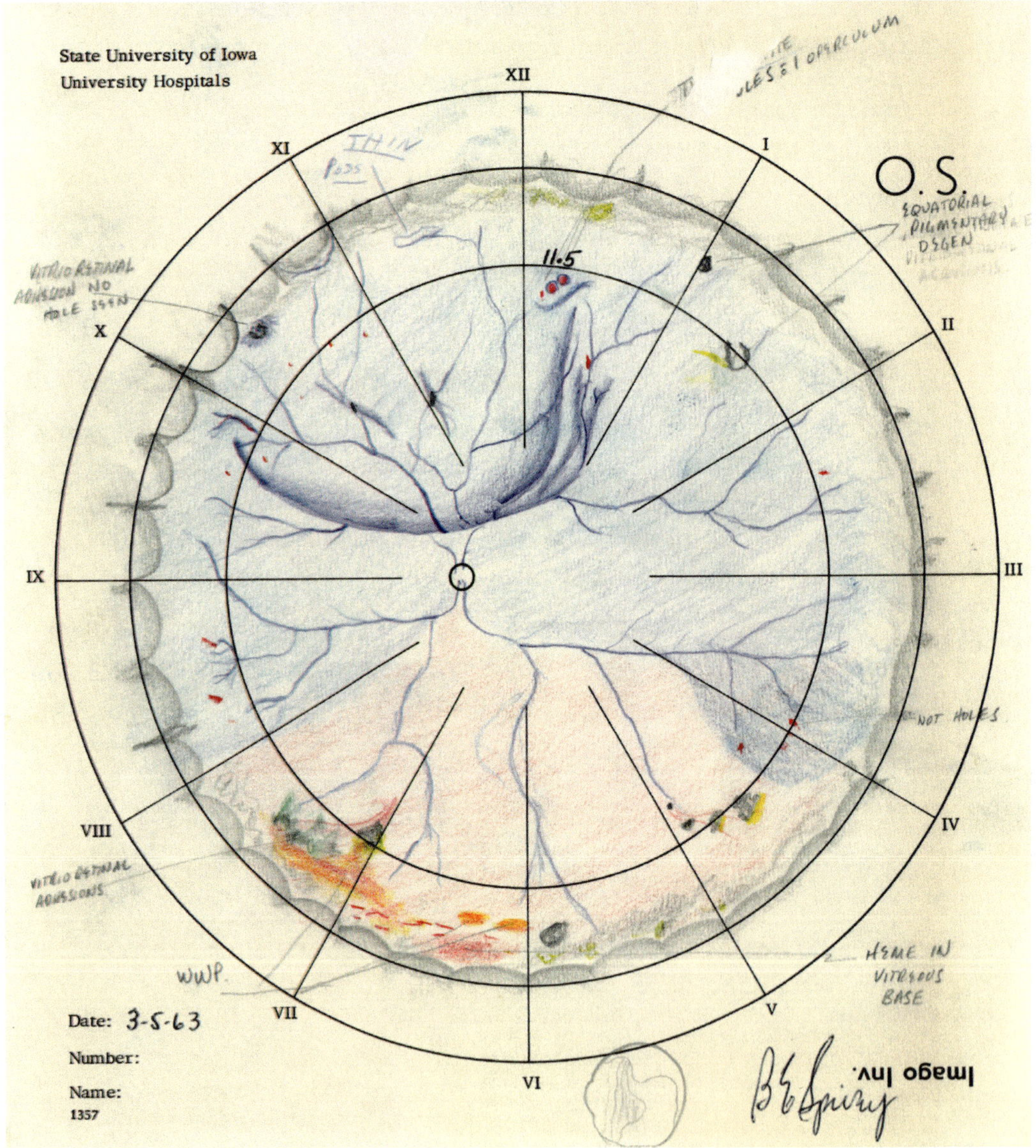

Artist Bruce E. Spivey
March 5, 1963

Diagnosis:
Retinal detachment and equatorial pigmentary degeneration, left eye, following cataract extraction six months earlier.

Artist H. Stanley Thompson
September 13, 1965

Diagnosis:
Retinal detachment, left eye.

"After the initial mess we made of drawings—just trying to do indirect and draw what one thinks he sees became an art project. Little did I know that Neal [Jepson] had art training until now! I did not, but I did do wood carvings as a kid, and the artistry of ophthalmology was one of the stimulating factors that drew me into the field."

David Fenske, MD (12/6/10)

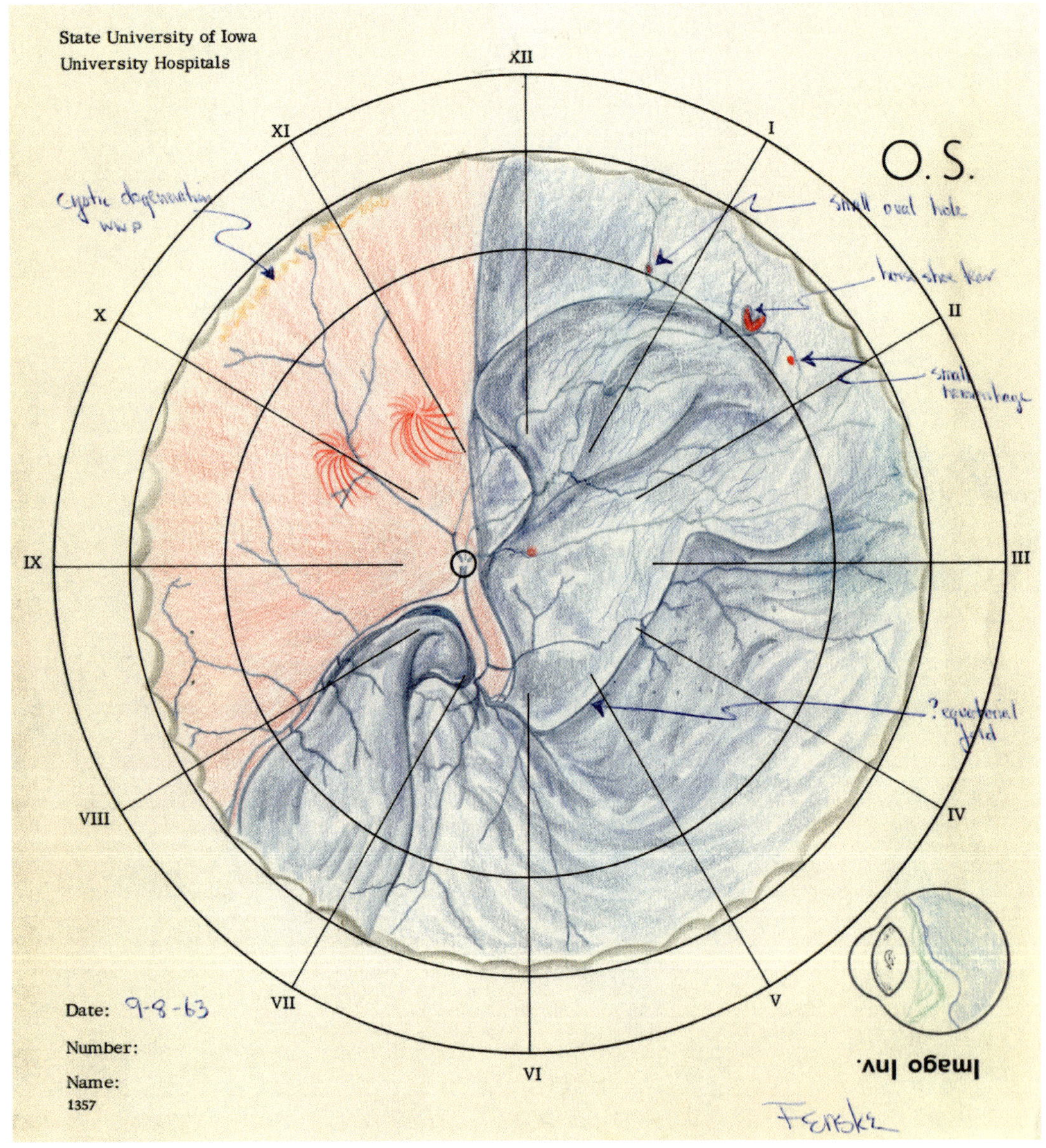

Artist David Fenske
September 8, 1963

Diagnosis:
Large bullous detachment,
left eye.

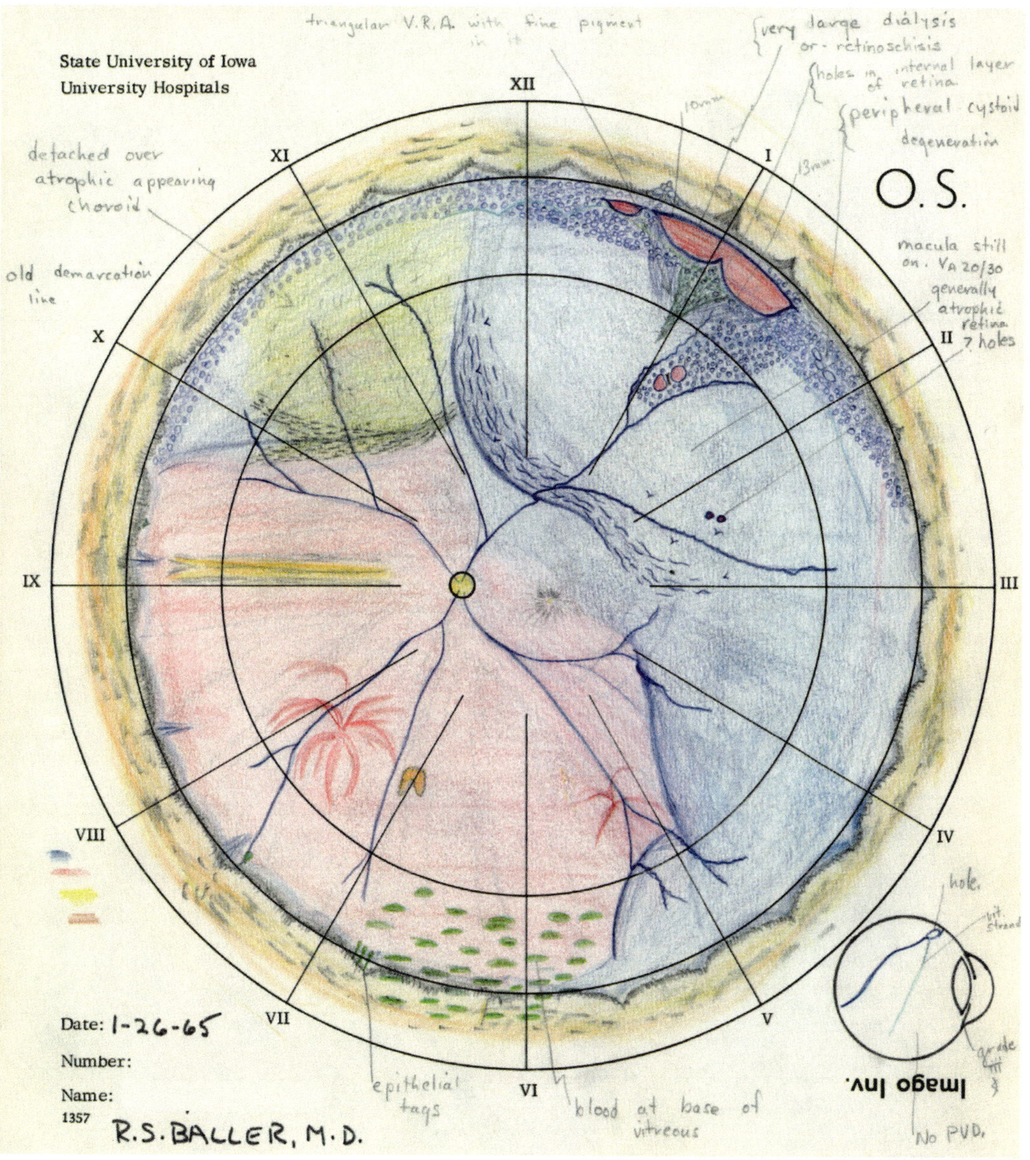

Artist Robert S. Baller
January 26, 1965

Diagnosis:
Retinal detachment, left eye.

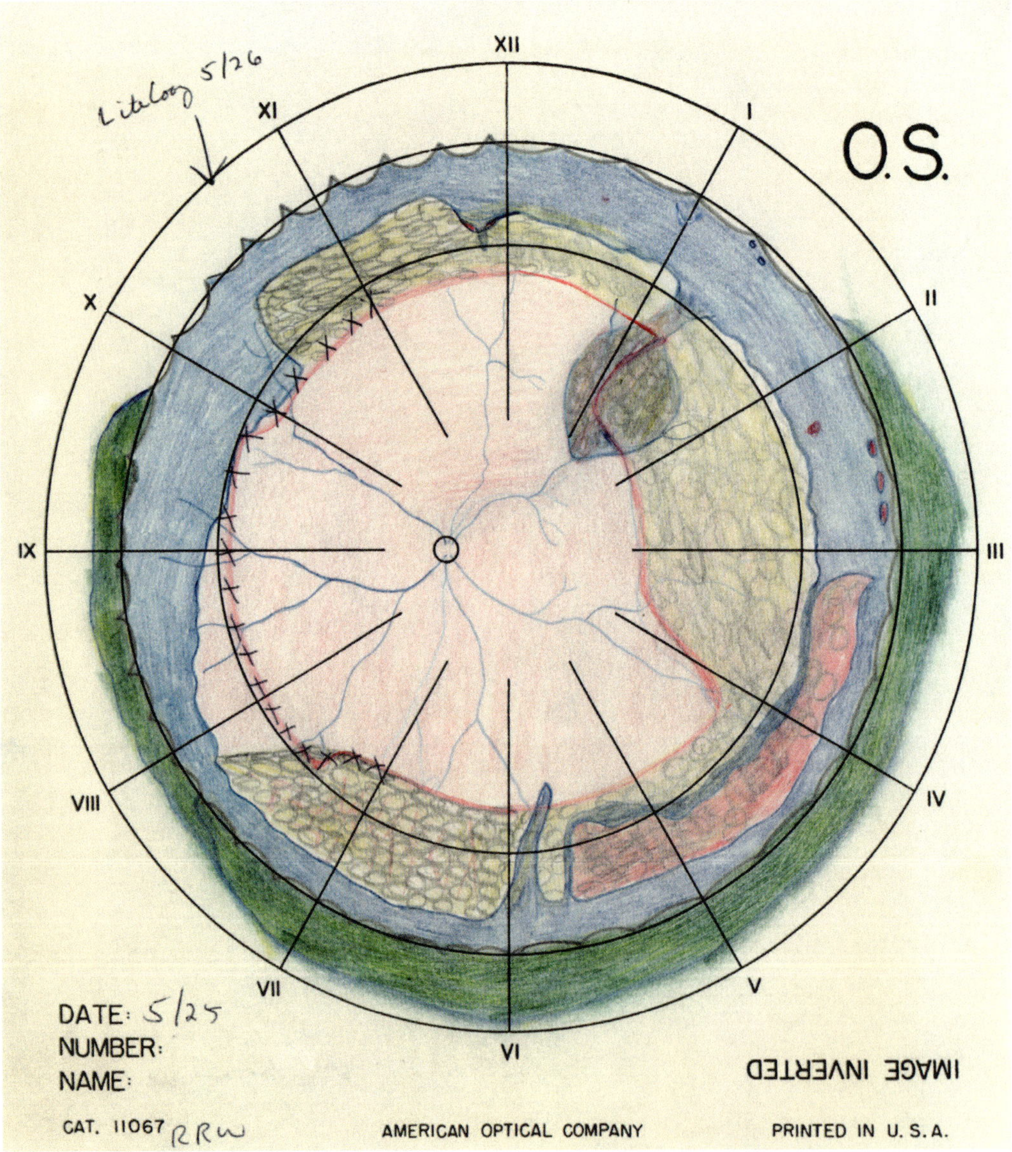

Artist Russell R. Widner
May 25, 1965

Diagnosis:
"Fluid dissecting from 12-degree tear nasally anterior to buckle. Severe vitreous traction is pulling ora off," left eye, following two scleral buckles and photocoagulation.

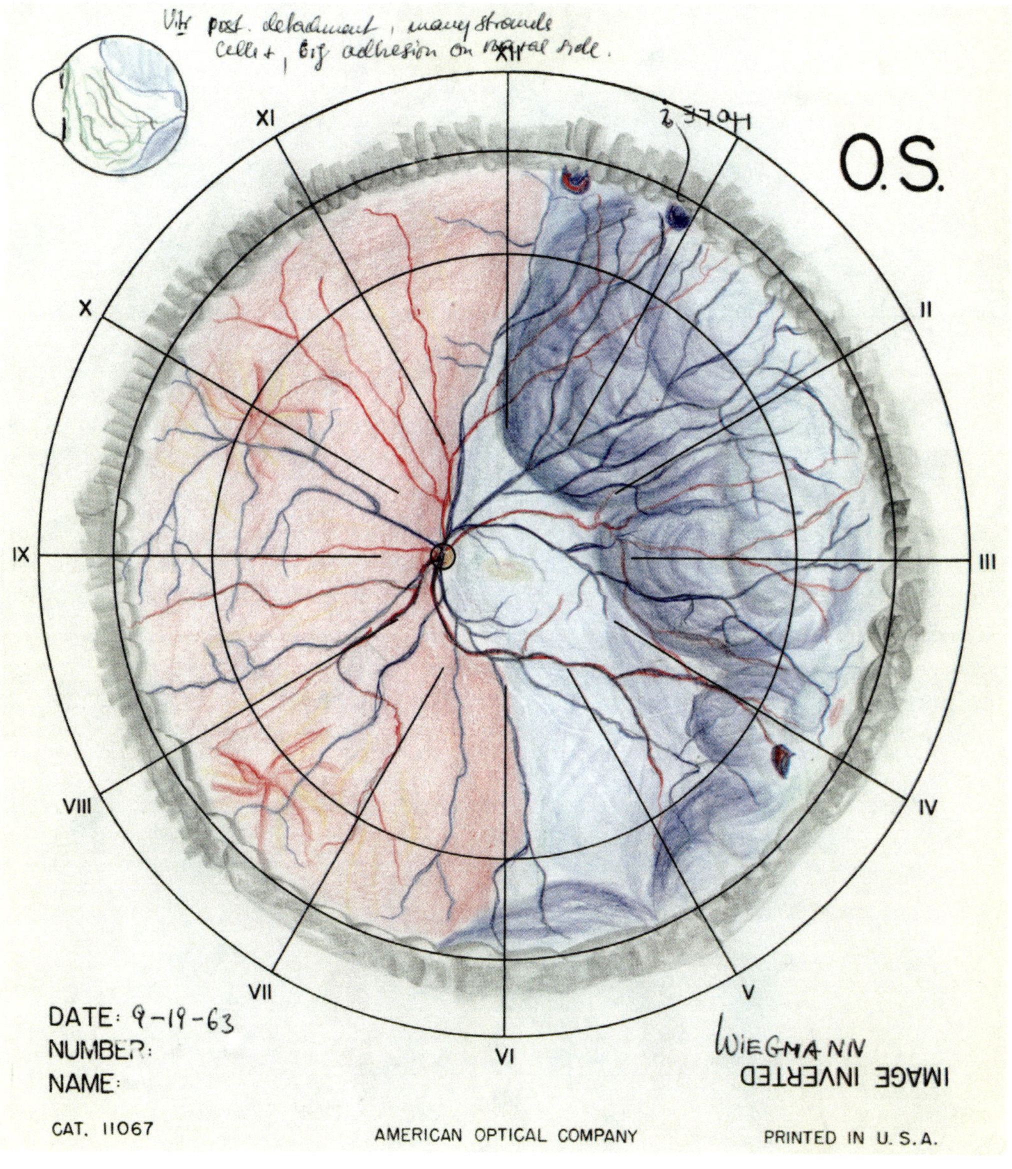

Artist Otto A. Wiegmann
September 19, 1963

Diagnosis:
Retinal detachment, left eye.

Dr. Watzke "belongs in the Pantheon of our profession."

T. Kim Krummenacher, MD (6/6/11)

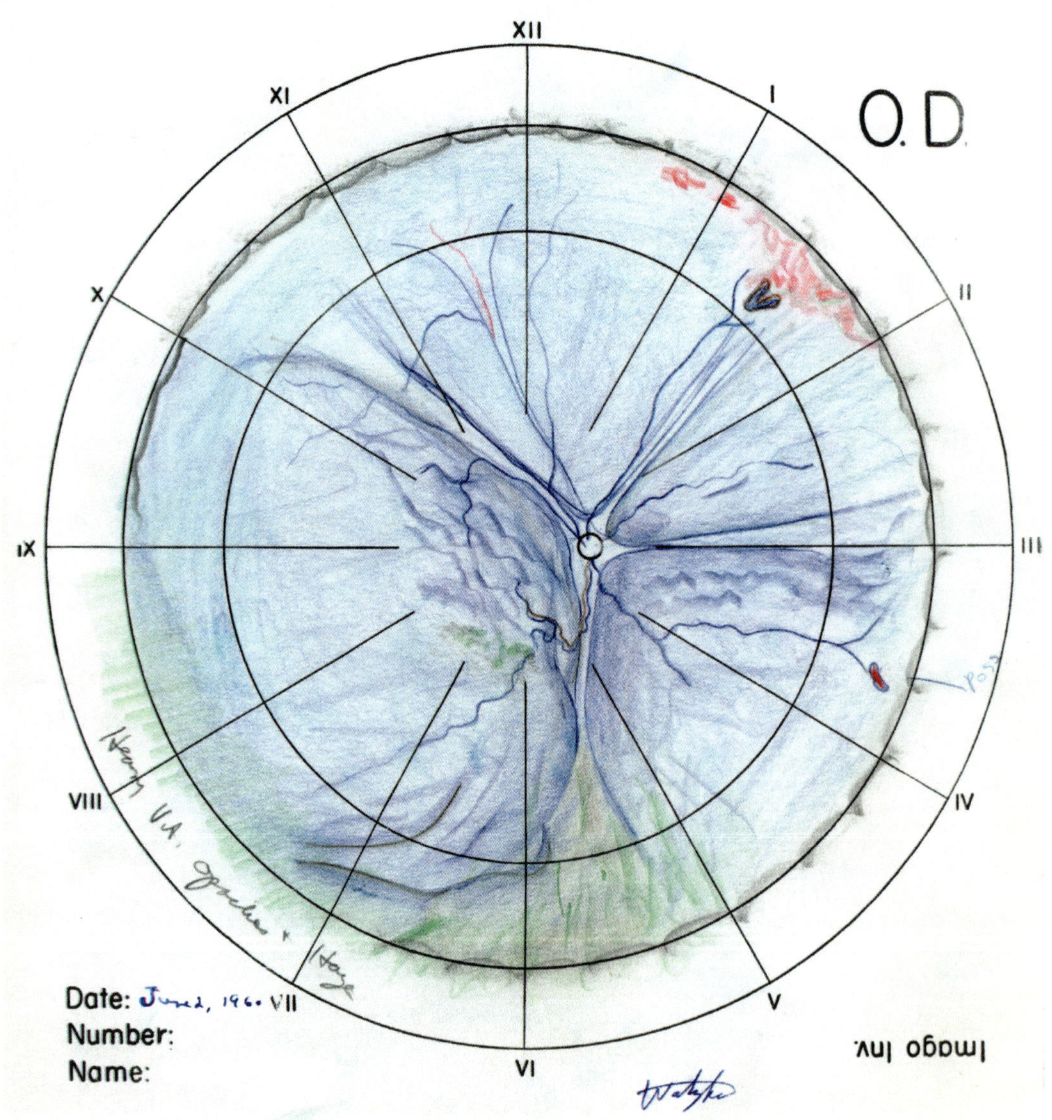

Artist Robert C. Watzke
June 2, 1960

Diagnosis:
Complete retinal detachment, right eye, following vitreous hemorrhage eight months earlier.

"I recall presenting the patient at morning rounds, and Dr. Watzke, who also examined the patient, commented that he was more impressed with the art work than the accuracy of the drawing."

Richard Schultz, MD (12/8/10)

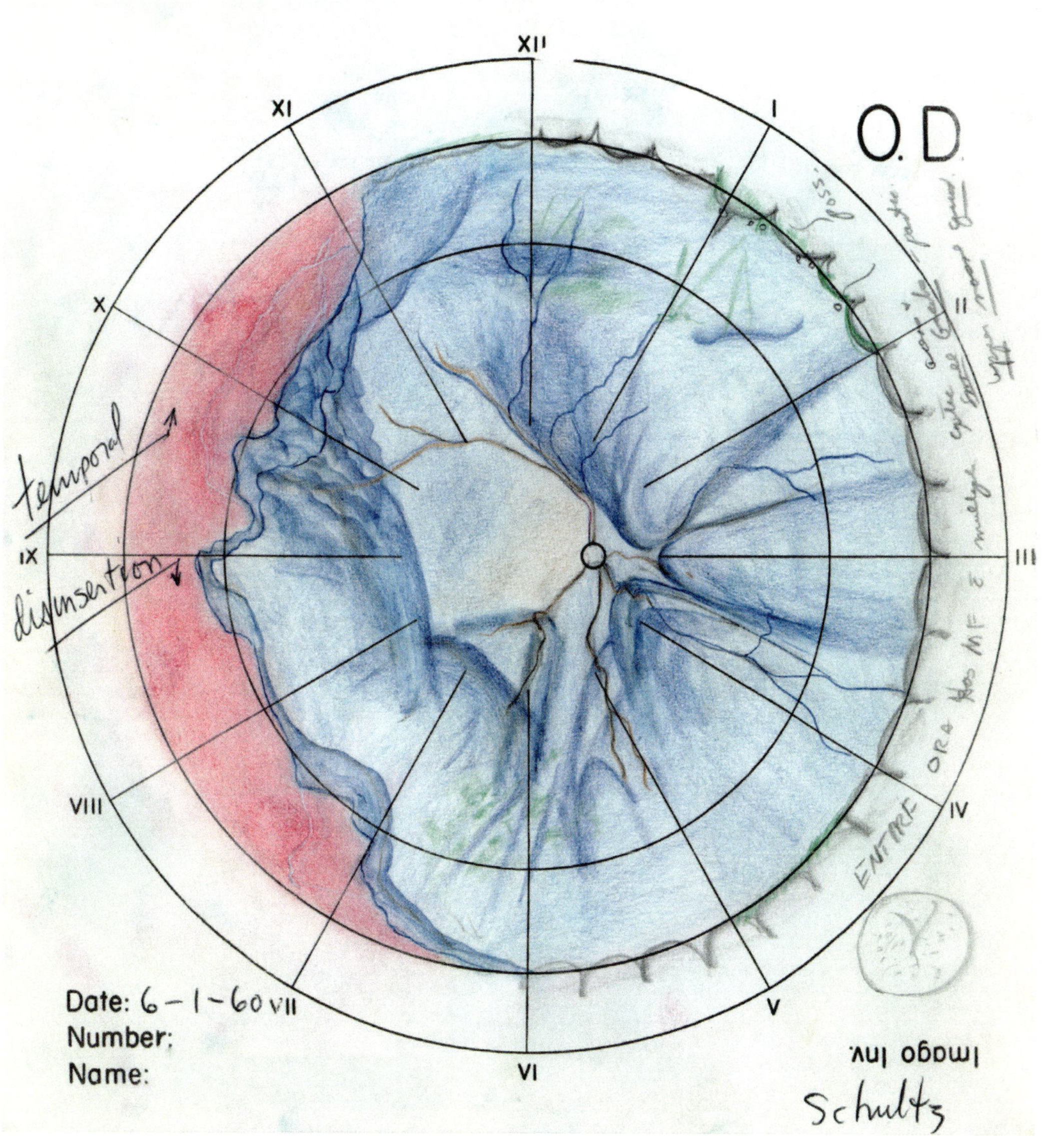

Artist Richard O. Schultz
June 1, 1960

Diagnosis:
Retinal detachment, right eye, after being kicked in the head by a cow.

After failing an art school entrance examination, "[Matisse] was eventually accepted as a student under Gustave Moreau, who encouraged individual creativity and stressed the value of simplification."

Michael Marmor, MD, and James Ravin, MD, authors of The Artist's Eyes

Henri Matisse

A Glimpse of Notre-Dame in the Late Afternoon (Notre-Dame, une fin d'après-midi)
1902, Oil on paper
Buffalo, Albright-Knox Art Gallery

Modern Expression

A more modern ethos hit the retina room in the 1970s. The period of descriptive publications of retinal pathology had passed, and academic studies came to emphasize case series and results of surgical variations. With the updated indirect ophthalmoscope early in its second decade and better handheld lenses with which to examine the peripheral retina, fundus drawing became more fixated on pathology, less on normal or unaffected areas, and more expressive of the artists' perceptions. The effects of self-expression, editorialization, and artistic interpretation became more evident in the documentation of disease. In the 1970s the fundus drawings at UIHC evolved from the realistic depictions of the retina and peripheral retina to this modern, sleek form—to the brighter colors and larger shapes reminiscent of the avant-garde movements of the early twentieth century, especially Fauvism and Orphism, with a hint of Art Deco.

By comparison, the visual features in art during the Modern period, specifically colors, lines, and shapes, are as varied as the trends and movements within it. Fauvism, for example, features flatness, bold or intense color often placed over white to heighten the light effect, and figurative yet simplified forms. Orphism, which is an offshoot of Cubism, features even more abstraction with the juxtaposition of colors, tones, and shapes or, as Orphist Robert Delaunay called them, "colored planes," creating the reality of light and depth that a focus on objects themselves cannot. In Art Deco, there is again abstraction and simplification; the art is imposing and streamlined, with intense colors, smooth surfaces, and a metallic sheen.

In real life, Modernism existed from the late 1800s through the 1960s and included various styles and

"...nature is no longer a subject for description but a pretext."

Robert Delaunay, artist

Robert Delaunay

Simultaneous Contrasts: Sun and Moon (Soleil, lune, simultané 2)
1912-1913, Oil on canvas
New York, The Museum of Modern Art

techniques, different degrees of abstraction, and experimentation. In retina life, a modern movement existed in the 1970s. It included the beginnings of a shift in focus—from the whole fundus to the pathology itself—and a temporary shift in medium to watercolor.

The innovation and self-expression of the modern movement are best seen in Dr. Vermillion's use of watercolors to depict the retina. "Watercolor is a lot faster," explained Dr. Vermillion (10/10/11), who had majored in fine art as well as in general science as an undergraduate. "And with my background, [watercolor] was just another medium. For me it was just a way to save time by blocking in large areas with color and blotting back to give dimension. Cross hatching is tedious."

Since "time was of the essence," Dr. Vermillion said, "[using watercolor] let me get out of there by 10 p.m." Then he added, "I also know [watercoloring] tickled Watzke when I would do it. [I was a] typical resident seeking favor." Dr. Vermillion's visual expressions in this volume started in 1978, continued in 1979, and culminated with this chapter's masterpiece, which was "drawn" on January 27, 1980, and features not only watercolors but the color vermilion.

"Modernist" fundus drawings, including Dr. Vermillion's, contain demarcated shapes or planes of color and feature brighter colors that shimmer. They have fewer minute details than in the first decade of the fundus drawing era and are fairly flat and smooth when the pathology allowed it. They also were increasingly abstract, albeit still figurative. The 1960s realistic drawings were figurative as well, which is a significant point when noting the variations in style or the evolution of the drawings at Iowa.

When someone sees Picasso's *Chat Saisissant un Oiseau (Cat Catching a Bird)* (1939) and notes that the painting really does show a cat eating a bird, the painting can be classified as figurative. It represents and somewhat resembles an actual object or scene. Similarly, the fundus drawings of the modern era, the 1970s in this context, are figurative.

Although the drawings become increasingly simplified and abstract after the 1960s, similar to a trend in Orphist paintings in the early twentieth century, it is important to remember that the drawings depict subject matter not readily recognizable to a lay audience because retinas are not usually seen in daily life except by ophthalmologists.

The reliance on form and color in Modern art and the sensations these expressions evoke portray the artistic freedom evident in the modernist retinal drawings, as does the narrowing of focus. "It was not until the '70s that increased experience with Indirect Ophthalmoscopy, the use of the Goldman lens and later more powerful 20–90 diopter lens provided greater detail be appreciated," explained Dr. Diamond (5/23/11). "I, in particular, attempted to use artistic license to enlarge the pathology to be more readily understood as aided by the advances in lenses. I focused on the pathology as I saw it in real time."

Possibly the 1970s were more about the drawings and veering off, challenging the short-lived tradition of drawing all of what the artist sees, as the artist sees it, and instead finding ways to focus on and accentuate the problem. This was a period of self-expression, of fixed focus; it contained a streamlined style of drawing soon to be refined or simplified even more.

Pablo Picasso

Cat Catching a Bird (Chat Saisissant un Oiseau)
1939, Oil on canvas
Paris, Picasso Museum

"[There was a] retina fellow a few years before me who employed professional airbrush equipment. Truly the 'belle époque' of this genre!"

T. Kim Krummenacher, MD (morning of 12/7/10)

"Exactement! Très jolie! That is your cover shot. Prettier than a Vargus girl!"

T. Kim Krummenacher, MD (afternoon of 12/7/10, after seeing this drawing)

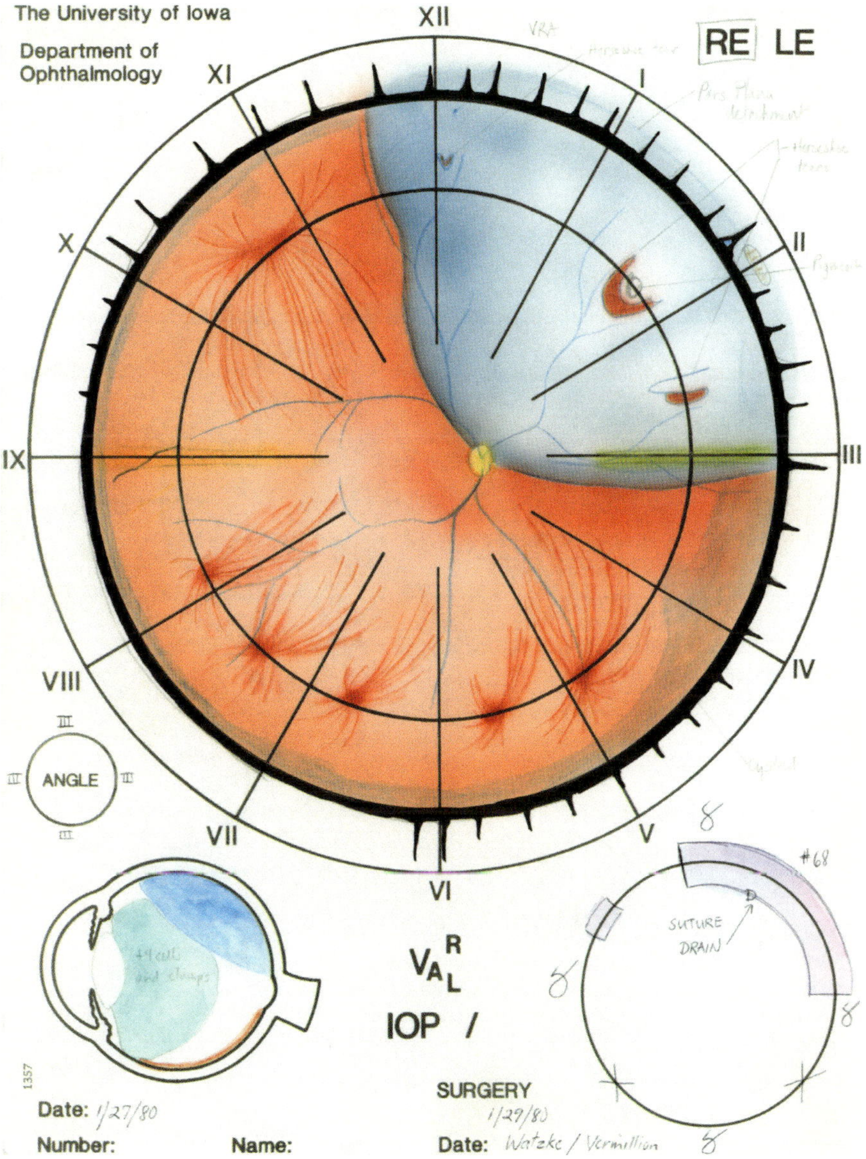

Artist Steven J. Vermillion
January 27, 1980

Diagnosis:
Retinal detachment, right eye,
with two flap tears.

"When Dr. Watzke and Dr. Burton came to check our drawings after we had been working at them for seemingly hours, it only took them a minute to check the drawings and make added corrections. I attributed part of that to the fact that I had already softened the eye enough to make their exam effortless."

Robert J. Thompson, MD (5/25/11)

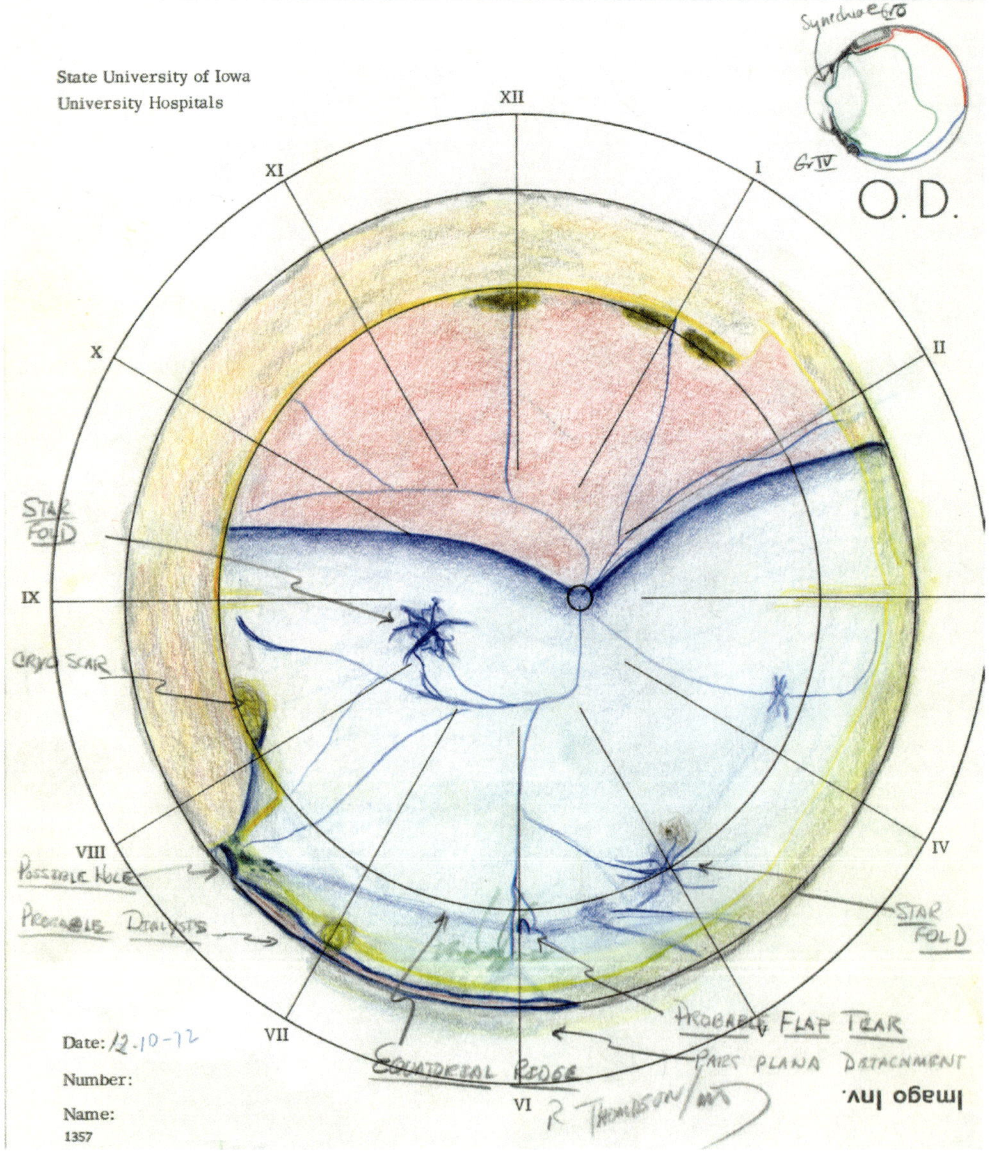

Artist Robert J. Thompson
December 10, 1972

Diagnosis:
Recurrent retinal detachment (with previous scleral buckle), right eye, following cataract extraction four years earlier.

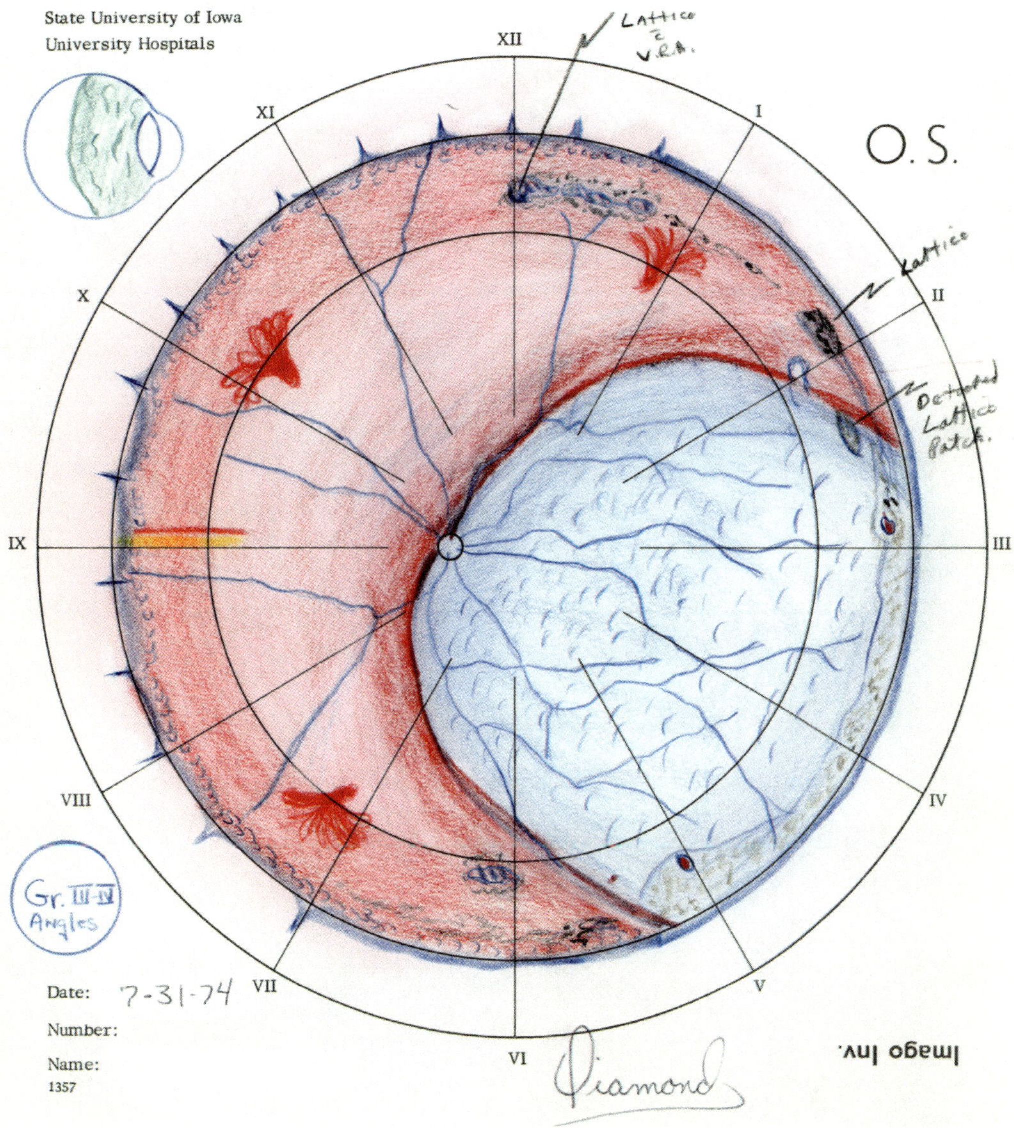

Artist James G. Diamond
July 31, 1974

Diagnosis:
Bullous retinal detachment with lattice and small holes, left eye.

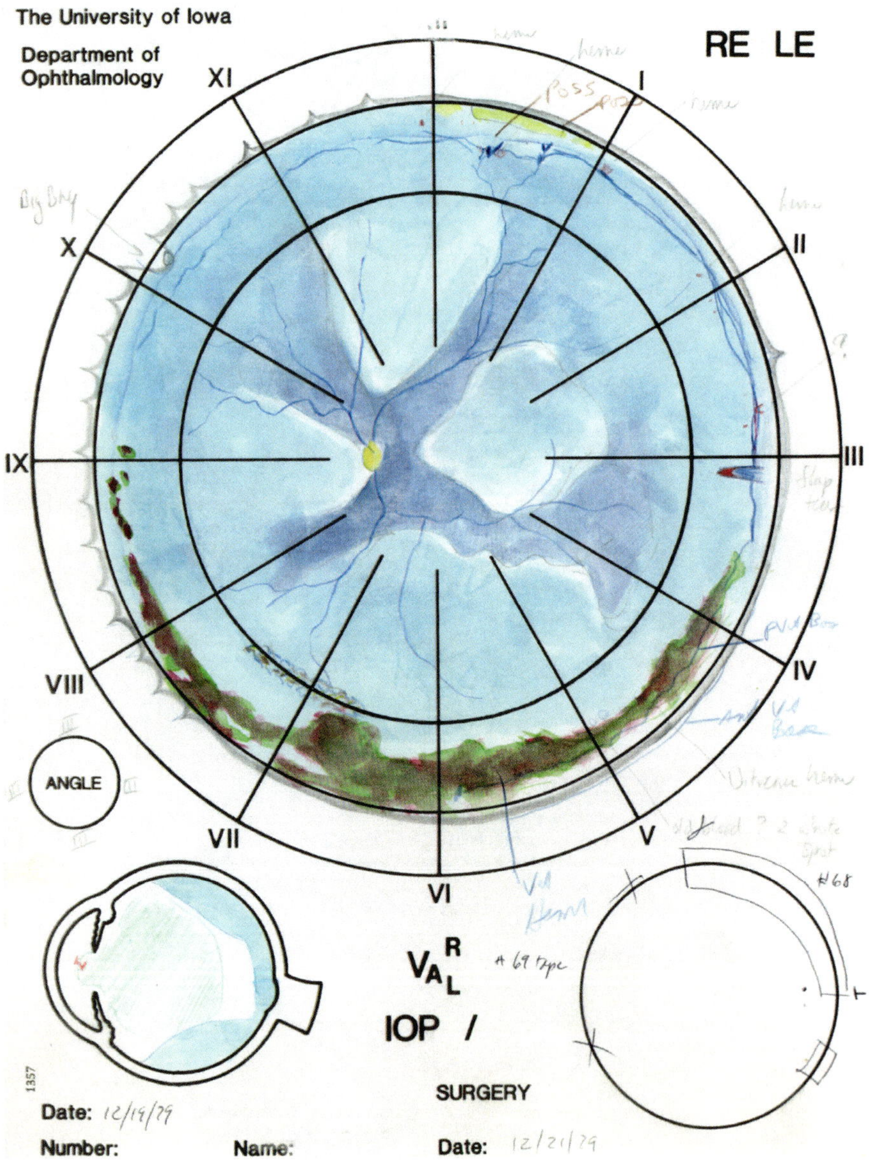

Artist Steven J. Vermillion
December 19, 1979

Diagnosis:
Total retinal detachment,
left eye, following cataract
extraction two months earlier

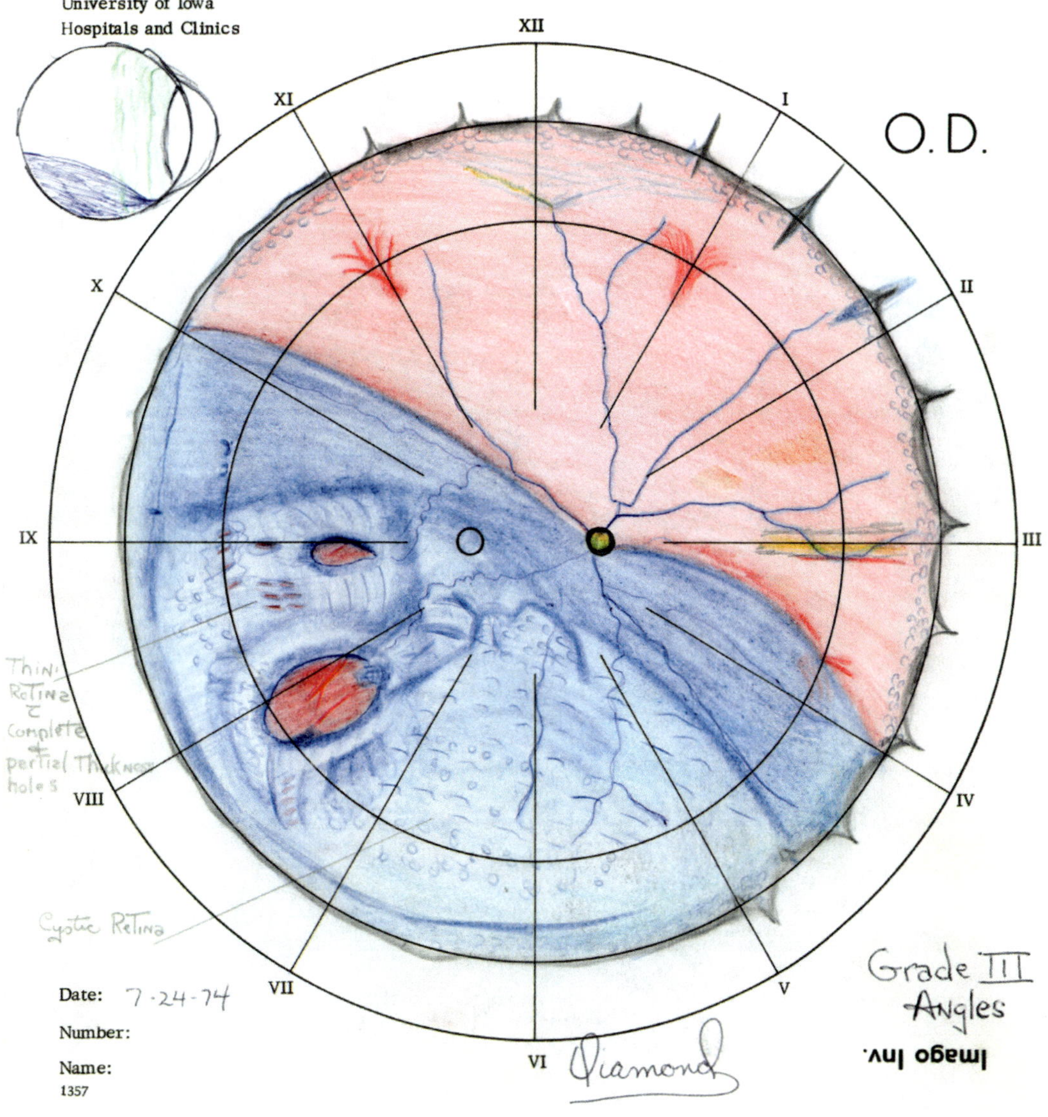

Artist James G. Diamond
July 24, 1974

Diagnosis:
Tractional retinal detachment secondary to trauma, right eye, with two large holes

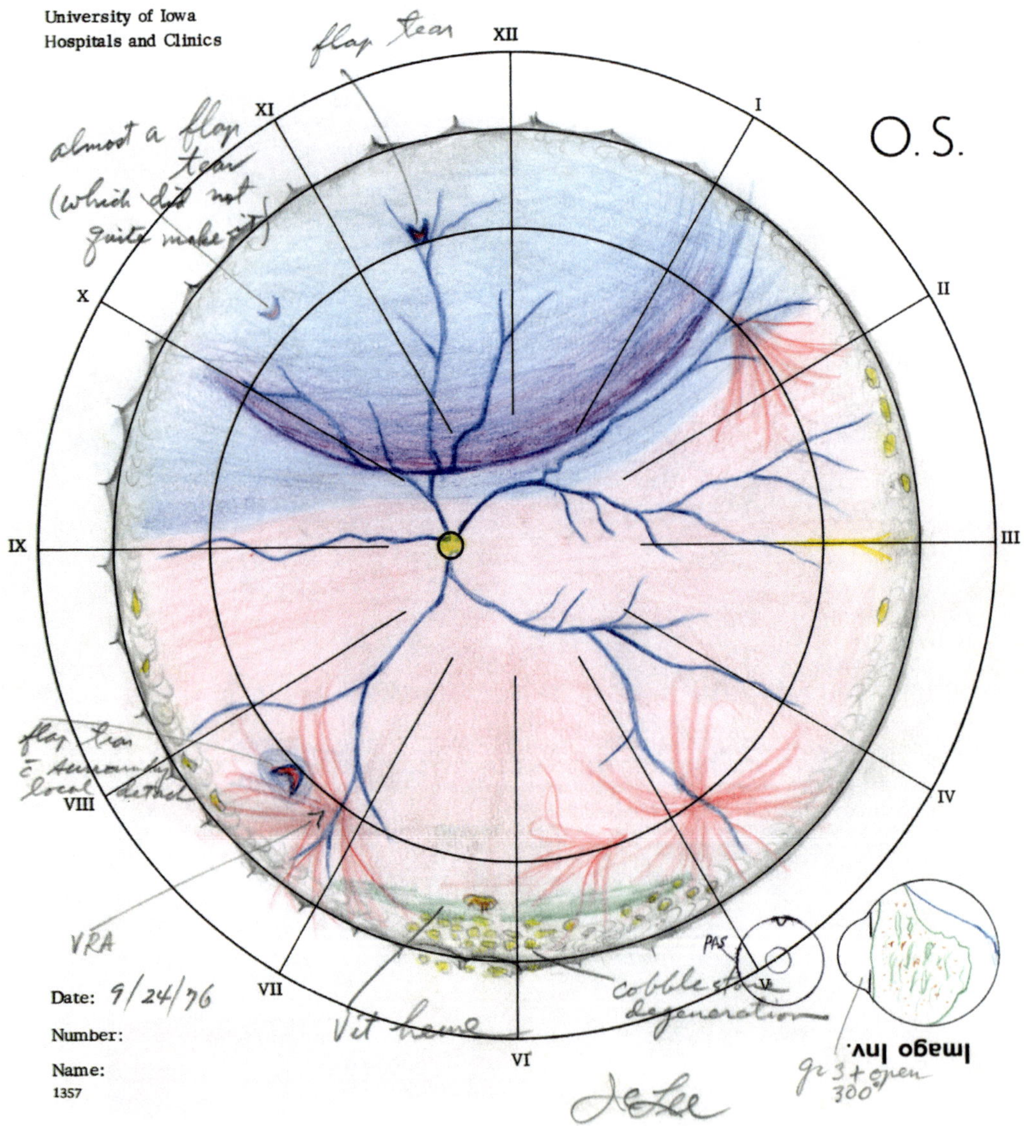

Artist John C. Lee
September 24, 1976

Diagnosis:
Retinal detachment, left eye, with flap tears; cataract extraction nine years earlier.

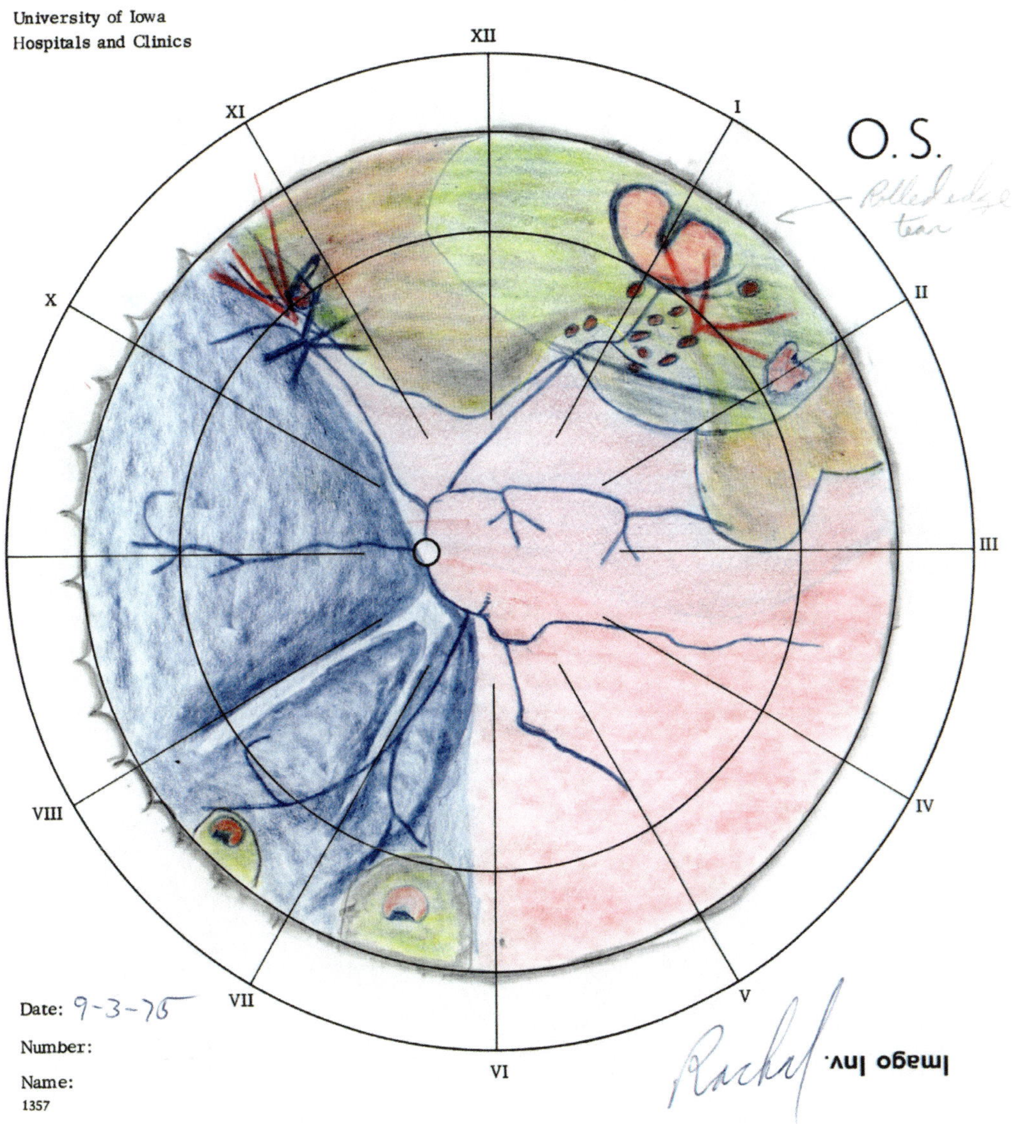

University of Iowa
Hospitals and Clinics

XII
I
II
III
IV
V
VI
VII
VIII
IX
X
XI

O.S.

Rolled edge tear

Date: 9-3-75
Number:
Name:
1357

Rachal
Imago Inv.

Artist William F. Rachal
September 3, 1975

Diagnosis:
Retinal detachment, left eye, following cataract extraction.

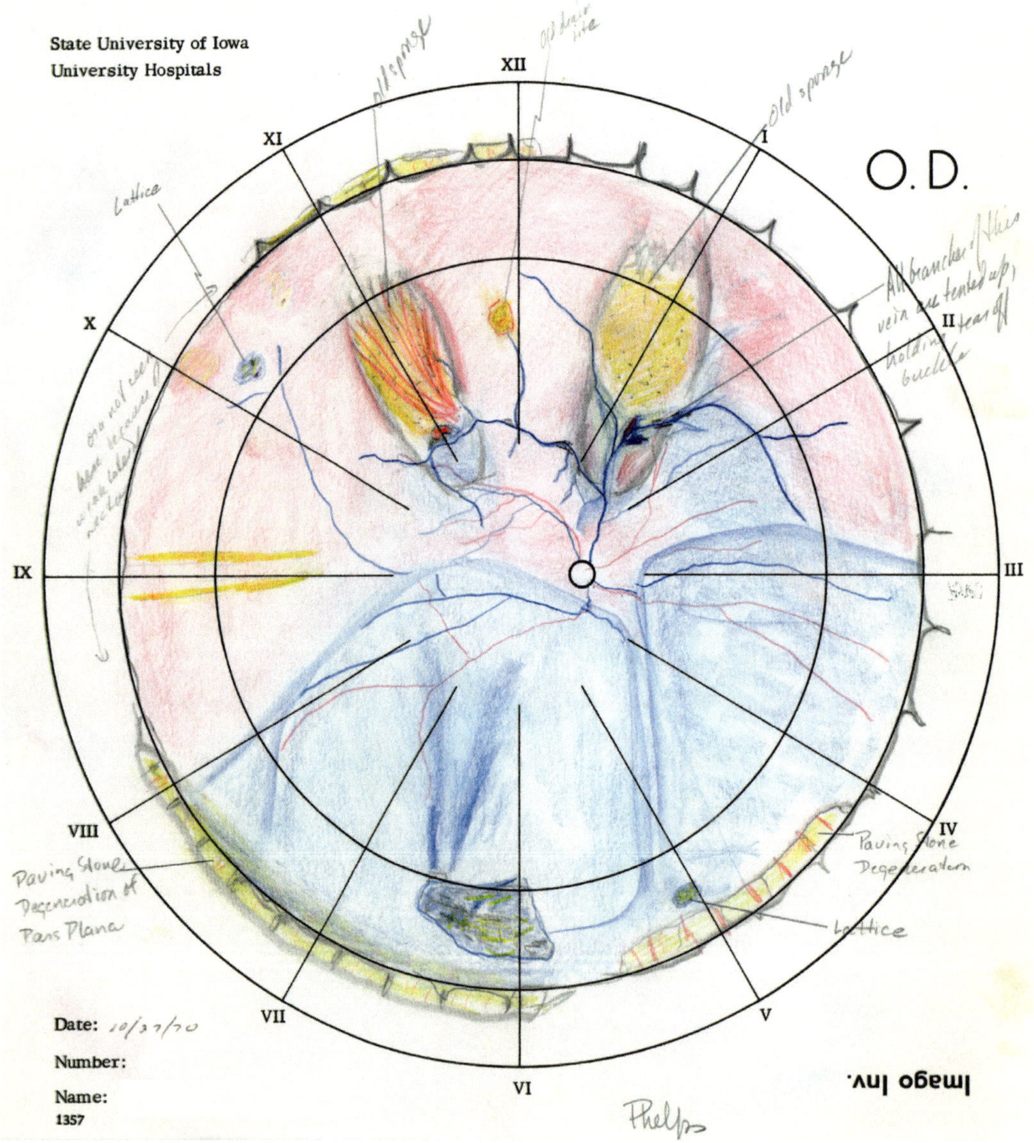

Artist Charles D. Phelps
October 27, 1970

Diagnosis:
Residual subretinal fluid following retinal reattachment with scleral buckling two weeks earlier, right eye.

"I didn't like filling in the background with the pencil, where it was supposed to be blue for detachment [or red for attached retina], so sometimes I took out my knife and shaved off the colored pencil. Then I smeared the color into the background...smudges."

Alex Smith, MD (9/9/11)

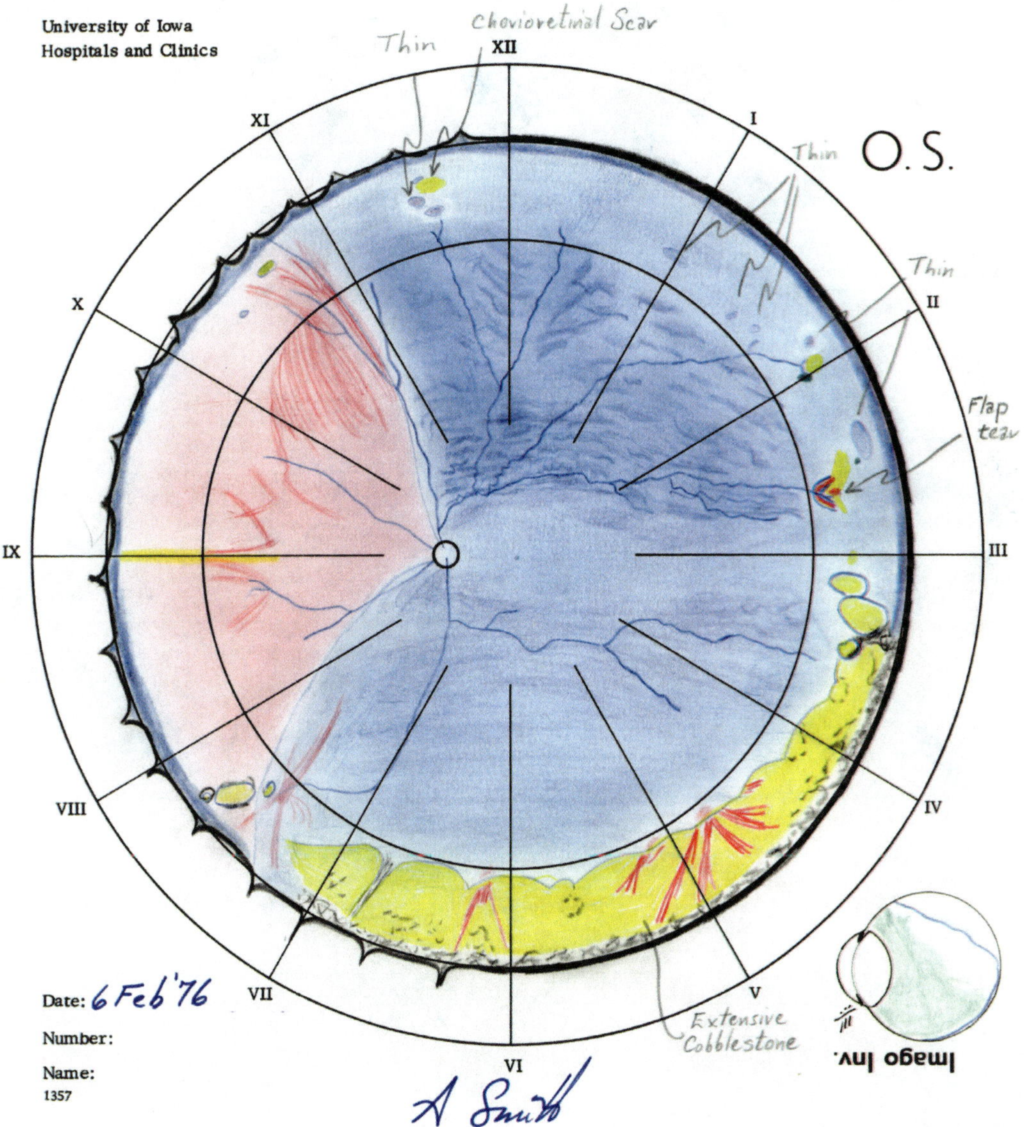

Artist Alexander G. Smith
February 6, 1976

Diagnosis:
Retinal detachment, left eye.

"In Redon's eyes, artists were the most important 'seers' of the beyond."

Norbert Wolf, art historian

Odilon Redon

The Eye Like a Strange Balloon Mounts Toward Infinity (L'Oeil, comme un ballon bizarre se dirige vers l'infini)
1882, Lithograph
New York, The Museum of Modern Art

Iconic to Symbolic Caricature

In the 1980s an even more abstracted mode of fundus representation permeated the majority of the drawings at UIHC, a mode characterized as "iconic to symbolic," thus having a range of decreased resemblance to the retina. In a wider context, Symbolism, known as "fundamentally antinaturalistic art[1]" or an attempt to "reveal the idea through figured representation[2]," is an art movement that, like the fundus drawings of the 1980s, describes a widely varied and visually abstracted universe. Instead of applying the terms iconic and symbolic to this section, we might easily have substituted cartoonism, cartoon art, line art, abstract line art, or even Surrealism, and for good reason. The drawings in it are varied; they contain different and possibly more personally meaningful features, unique when compared to other periods and unique within their own.

All of the drawings could be described as having simplified form, but they are not simple. These symbolist drawings are much more nonfigurative, however, when compared to the realist and the modernist drawings. Lines, dots, stick figures, cartoon creatures, and what appears to be an exploded cactus add a touch of playfulness or whimsy not seen in the previous periods at UIHC.

While art classified as Symbolism is antinaturalistic in both content and form, the symbolic fundus drawings merely emphasize form over content or accentuate form and subordinate some content in order to "illustrate a mood[3]," with mood being the pervading impression of a particular artist. Dr. Dreyer, for example, began explaining how he "experimented with a broad variety of techniques trying to convey, not realism, but...lots of very strong—" Then he stopped and broke into an example to illustrate the anti- or nonrealistic style he used (3/1/11). "In the '80s," he continued later, "we were intent on finding every last bit of pathology. [We asked,] 'What's the key?' 'What caused this retinal detachment?' 'What's the problem?' 'How do you portray it?'" This approach was "much more interesting" than drawing the retina realistically, he added.

Dr. Dreyer's mental dances affirm symbolism, the art of an idea—inasmuch as fundus drawing can express an idea rather than exclusively document a clinical condition. Truly, any of the three sections in this chapter or any of the other chapters could be titled "Symbolism" because the drawings are, at the most basic level, iconography: an actual code dictates color selection.

None of the drawings in this book, then, is totally realistic. Retinas are red. There is no blue or yellow retina. Vitreous hemorrhages are certainly not green; neither are translucent membranes. But these features are green on a fundus chart. Real lattice degeneration, on the other hand, may in a way resemble a lattice on a window or a fence. However, the crisscrossed lines representing retinal lattice are neither a structural nor an ornamental enhancement. Retinal lattice signifies deterioration; the lines are symbols, as are the parallel lines in retinoschisis (a splitting of the retinal layers) and the small round, often conjoined yellow circles of paving stone or cobblestone degeneration—circles that also sometimes contain parallel lines when drawn.

Even though symbolism was necessarily present throughout the entire retinal drawing era to communicate conditions, it is most pronounced in the drawings of the 1980s when the artists' ideas of what was key superseded realistic depictions, and the artists' decorative renditions evoked more than mere awe. The iconic to symbolic drawings induce giggles or, at the very least, smiles.

Distinct from the dark period in symbolist painting, iconic to symbolic fundus art from Iowa features lighter colors and lines of colored shading. Like symbolist art overall, however, the fundus art of the '80s features possibly the most personal signs of the era. For example, after glancing at drawings by Drs. Coonan and Caskey (July 1987 and December 1986 in this section), Dr. Watzke wrote, "To me these drawings are unclear; was there a total detachment in each with the retina contracted into a fibrous ball at the disc? If so, why are retinal vessels drawn?" (2/28/11) It seems symbolist drawings induced criticism as well as smiles.

"I think that as time went on and surgery became more routine with many different options," Dr. Watzke added, "the drawings became more 'cartoonish.'"

That might be. Like the symbolist art of Odilon Redon or Gustave Moreau, and like the cartoon caricatures of Disney and Warner Brothers animator Thornton "T." Hee, the symbolist retinal art represents the artist's ideas more than the artist's literal view. T. Hee once said that caricature is "the art of observation and study [4]," and the same could be said about fundus drawing. "A caricature is a mental image—an impression," he said. "Your eye makes no impression at all. Your mind makes the impression."

This applies to formal fundus drawing in its final decade as well.

1. Reinhold Heller, "Concerning Symbolism and the Structure of Surface," Art Journal 45:2 (June 1985): 146-153.

2. G.-Albert Aurier (as cited in Heller, 1985, p. 146)

3. Amy Dempsey, Styles, Schools and Movements: The Essential Encyclopaedic Guide to Modern Art (London: Thames & Hudson, 2010), 43.

4. Amid Amidi, Cartoon Brew, http://www.cartoonbrew.com/tag/t-hee, (April 6, 2011).

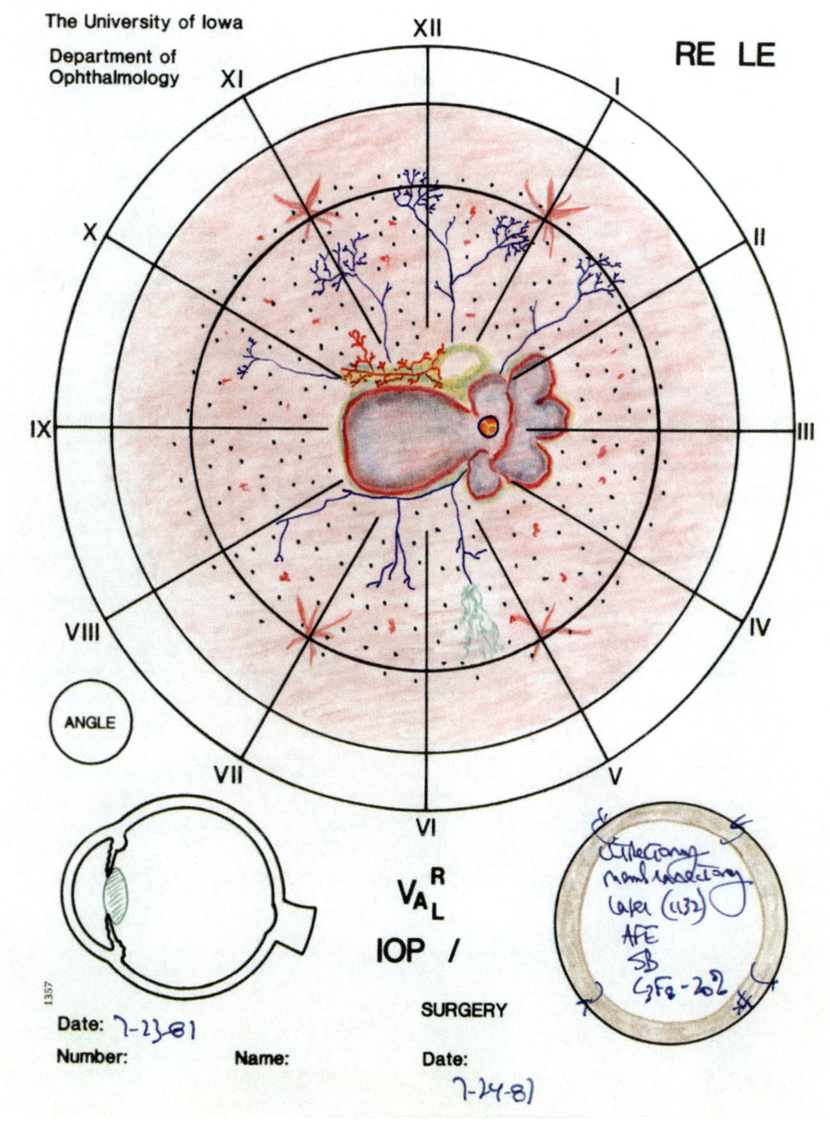

The University of Iowa
Department of Ophthalmology

RE LE

XII I II III IV V VI VII VIII IX X XI

ANGLE

VA R L
IOP /

SURGERY

Date:
Number:
Name:
Date:

Artist Patrick Coonan
July 23, 1987

Diagnosis:
Subhyaloid vitreous hemorrhage and tractional retinal detachment, right eye. Proliferative diabetic retinopathy.

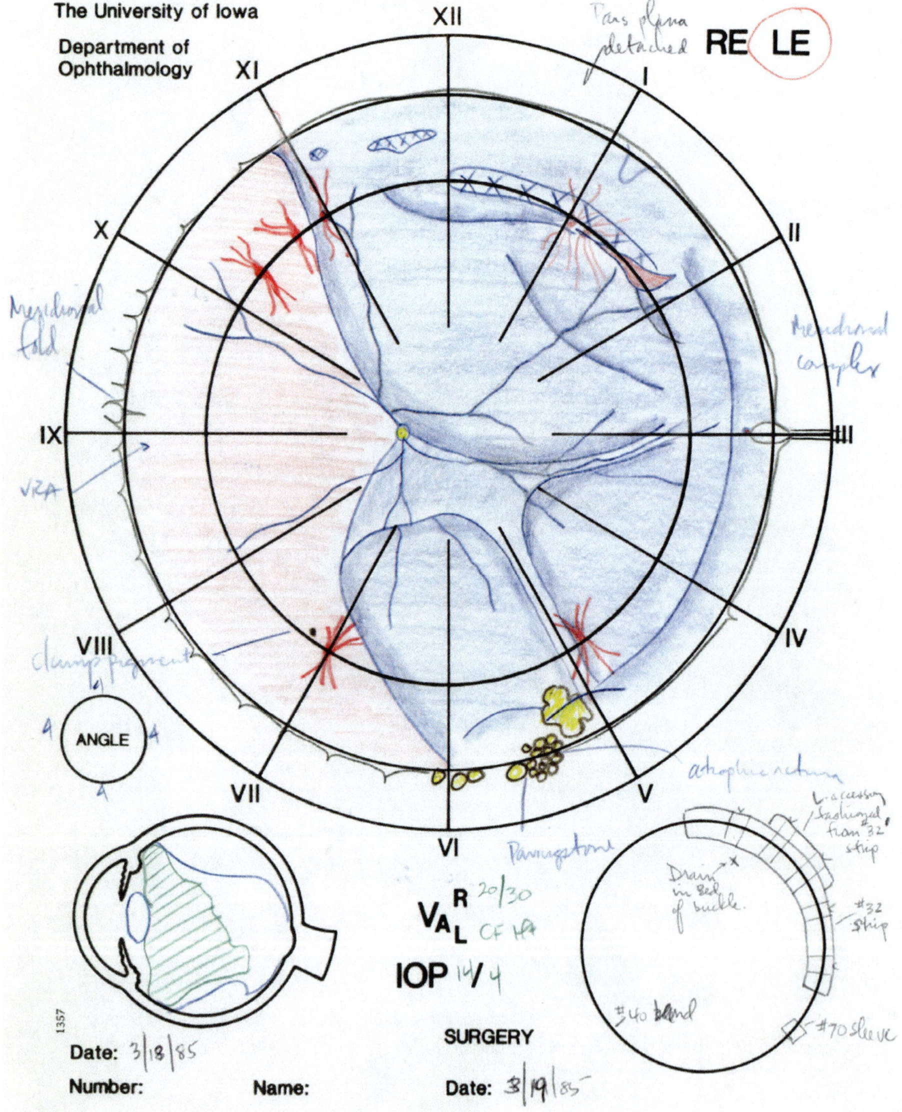

Artist Dennis P. Han
March 18, 1985

Diagnosis:
Superotemporal retinal detachment and lattice degeneration, left eye.

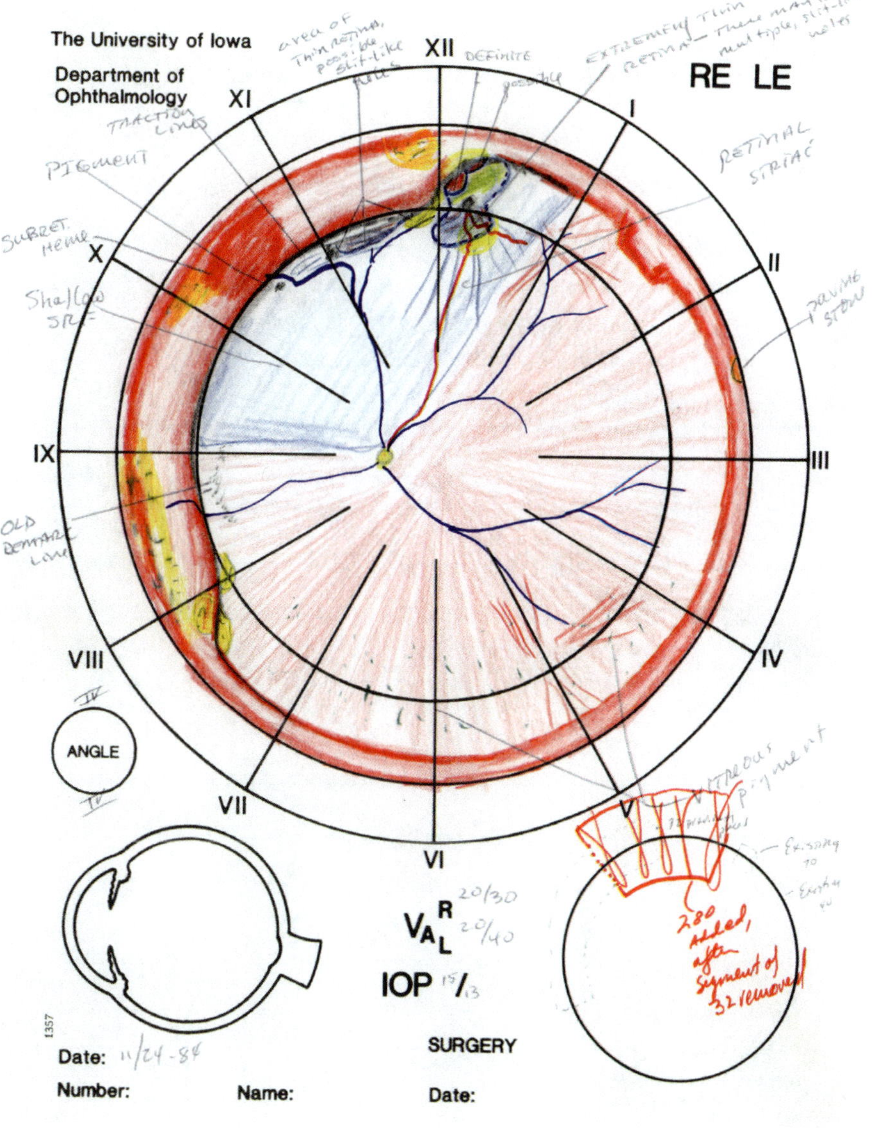

Artist Richard F. Dreyer
November 24, 1984

Diagnosis:
Recurrent retinal detachment, left eye, following scleral buckling one month earlier and cataract extraction five years earlier.

"It really is a bit sad to think that retinal drawings such as these are no longer necessary to perform as a retinal specialist with the advent of newer imaging technologies, but such is the price of progress."

Patrick Caskey, MD (4/20/11)

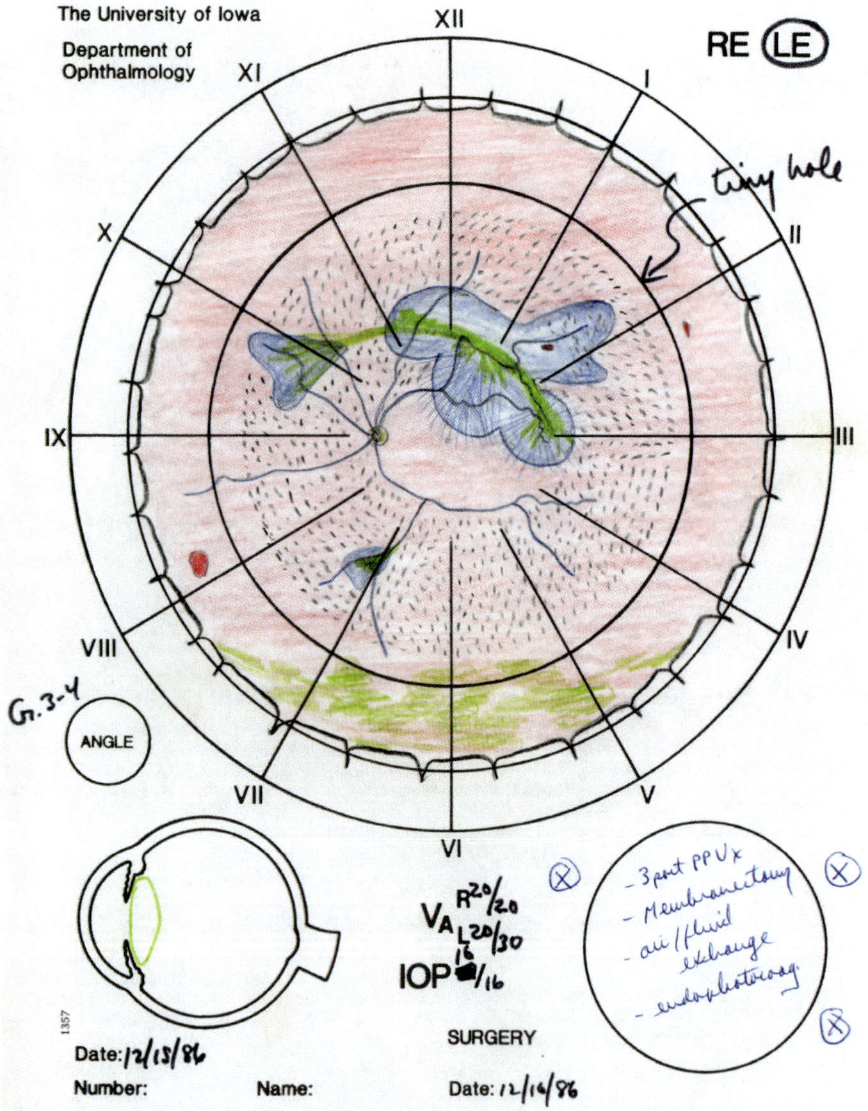

Artist Patrick J. Caskey
December 15, 1986

Diagnosis:
Tractional retinal detachment, left eye. Proliferative diabetic retinopathy.

"This is one of my early, cruder works from my 'blue period.' ... My later works I am certain are more arresting. ... Please include this drawing in the chapter entitled 'First Attempts.'"

T. Kim Krummenacher, MD (12/6/10)

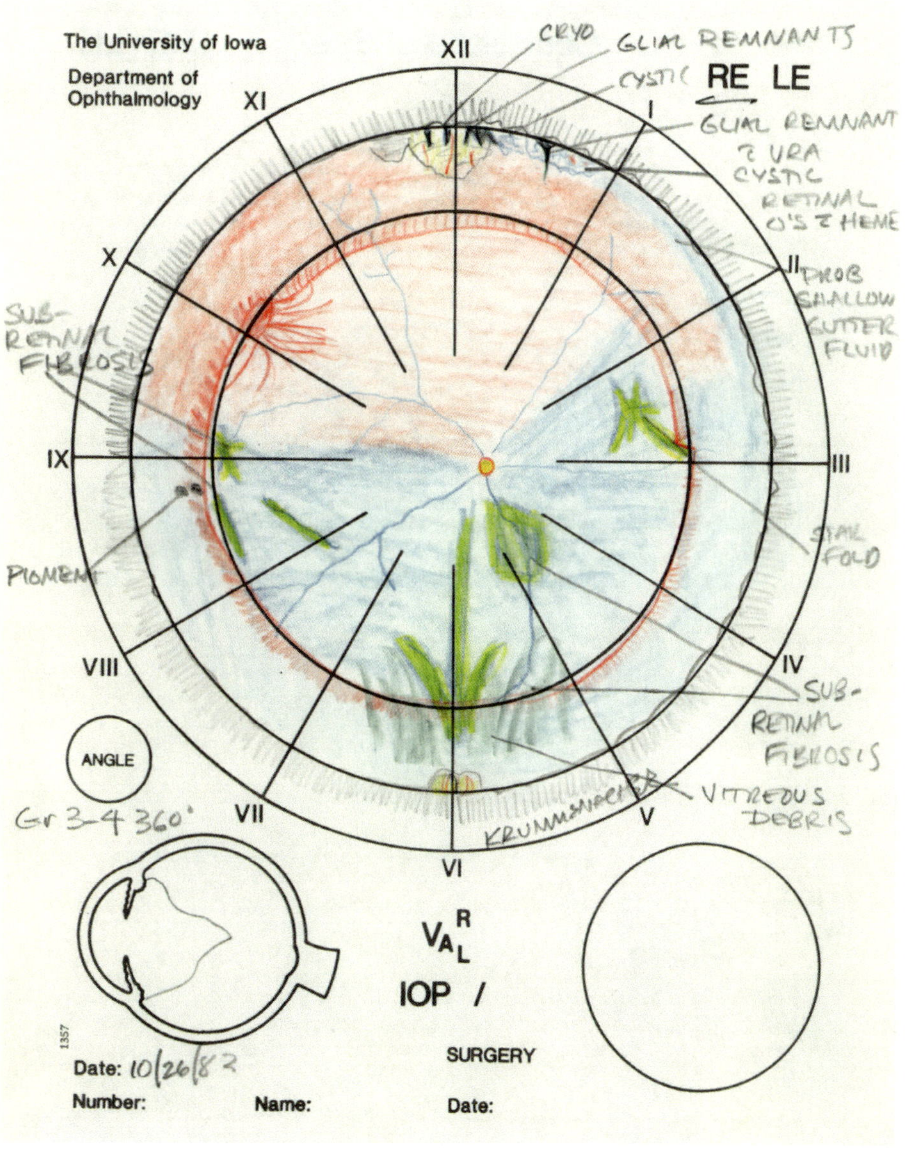

Artist T. Kim Krummenacher
October 26, 1983

Diagnosis:
Recurrent retinal detachment following scleral buckle one month earlier and cataract extraction with massive vitreous loss one year earlier, right eye; periretinal fibrous proliferation; and interstitial keratitis.

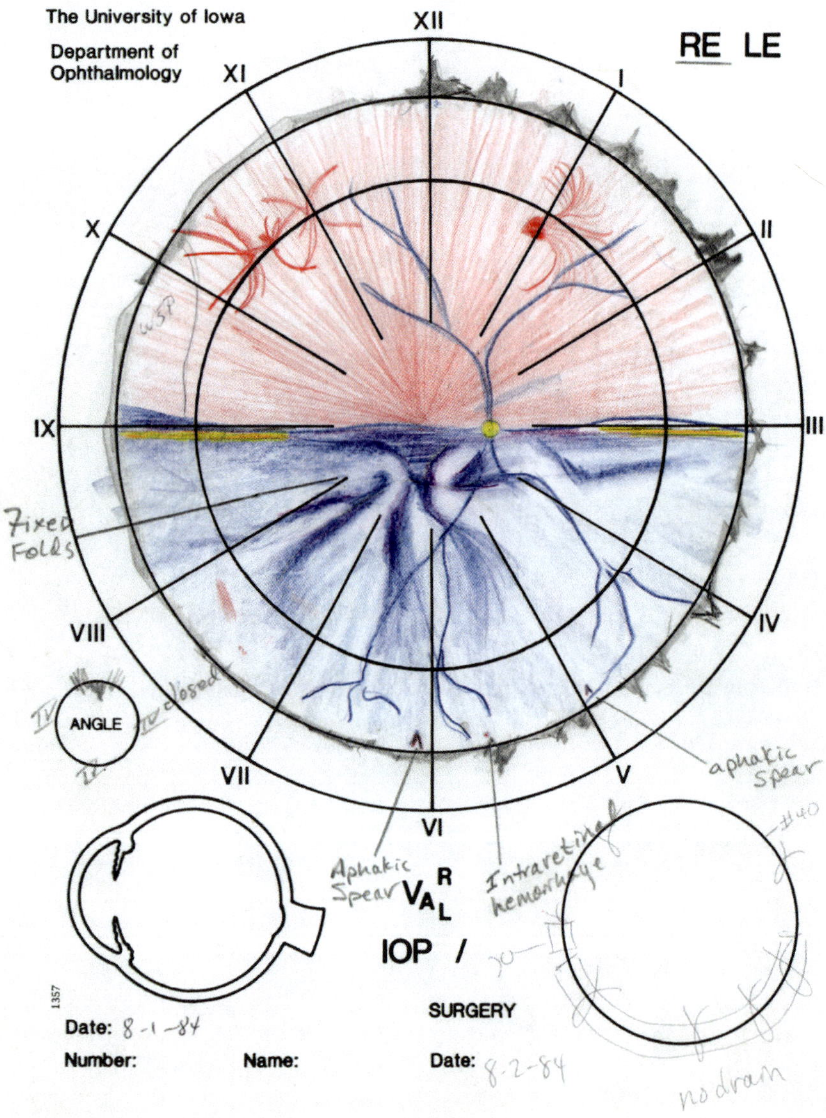

The University of Iowa
Department of Ophthalmology

RE LE

XII I II III IV V VI VII VIII IX X XI

Fixed Folds

ANGLE

aphakic Spear

Aphakic Spear

Intraretinal hemorrhage

VA_L^R

IOP /

Date: 8-1-84
Number:
Name:

SURGERY
Date: 8-2-84

no drain

Artist Richard F. Dreyer
August 1, 1984

Diagnosis:
Rhegmatogenous retinal detachment, right eye, after being hit in the head with a steel rod eight months earlier and after having an extraction of a congential cataract twenty years earlier.

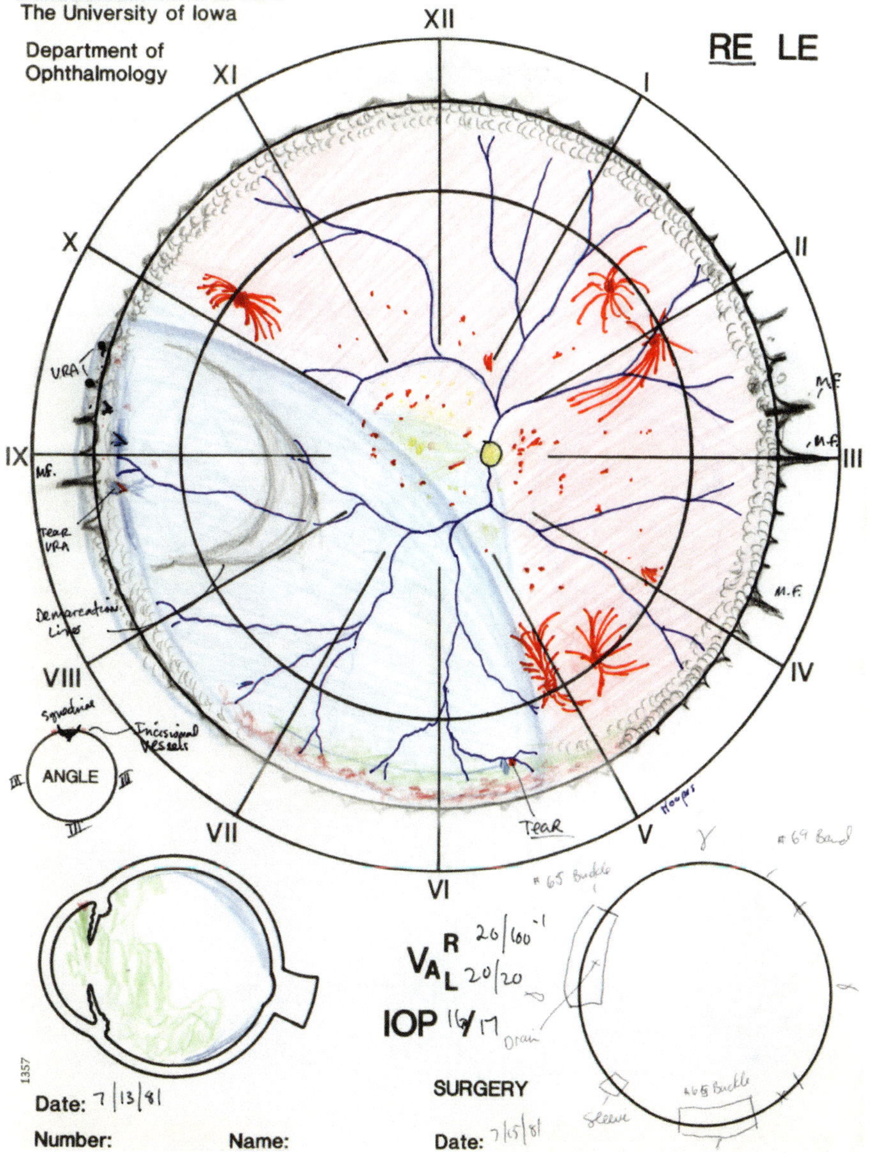

Artist Phillip C. Hoopes
July 13, 1981

Diagnosis:
Retinal detachment, right eye; diabetic retinopathy; cataract extraction eight years earlier.

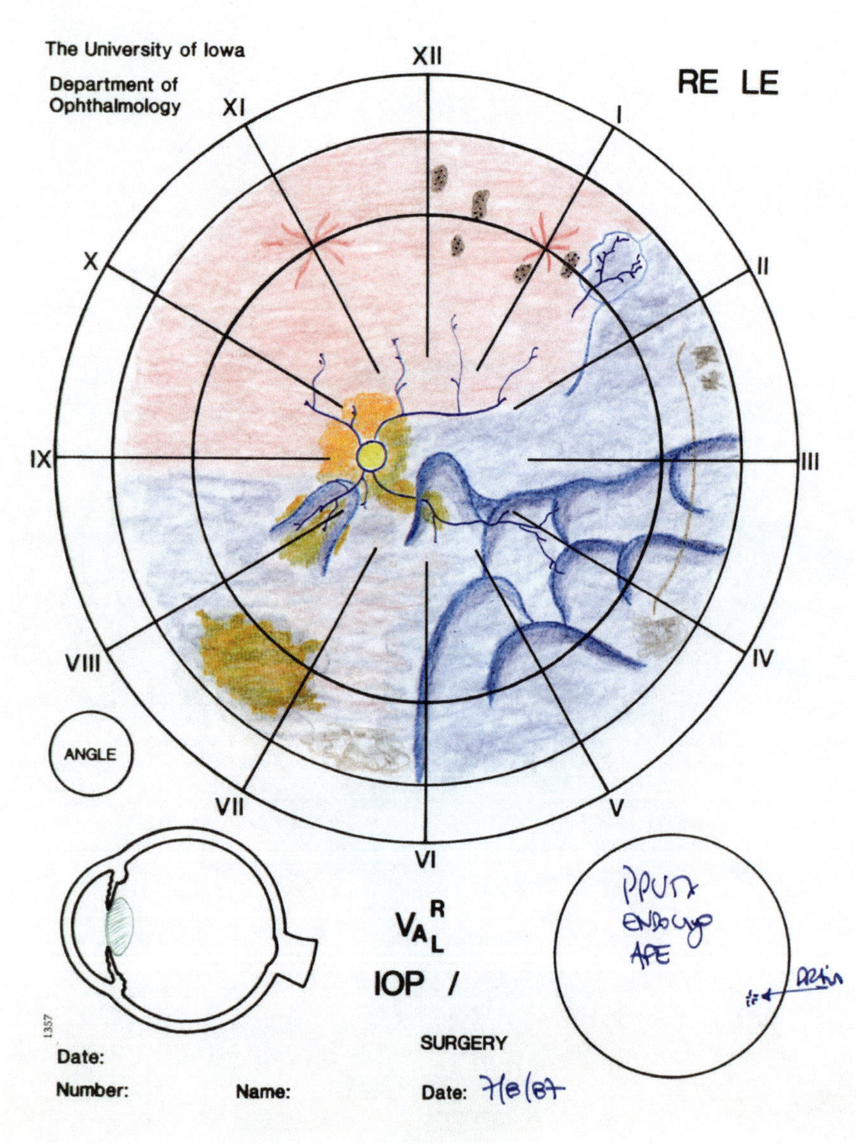

The University of Iowa
Department of Ophthalmology

RE LE

XII I II III IV V VI VII VIII IX X XI

ANGLE

VA R L

IOP /

SURGERY

Date:

Number: Name:

Date: 7/8/87

Artist Patrick Coonan
July 8, 1987

Diagnosis:
Bullous exudative detachment, left eye, secondary to Coats' disease. History of numerous treatments including photo-, cryo-, and lasercoagulation and scleral buckling.

"A little baroque, isn't it?"

Sam Farmer, MD (1/3/11)

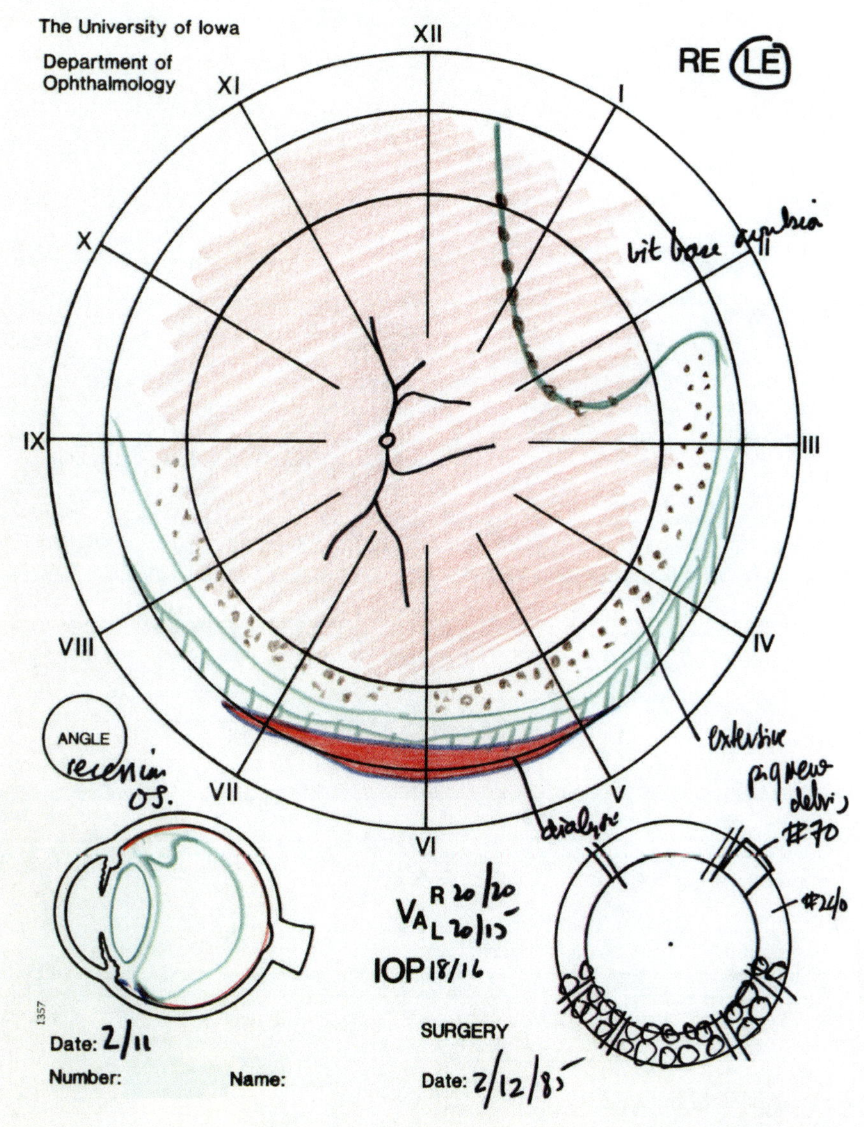

Artist Samuel G. Farmer
February 11, 1985

Diagnosis:
Traumatic retinal dialysis, left eye.

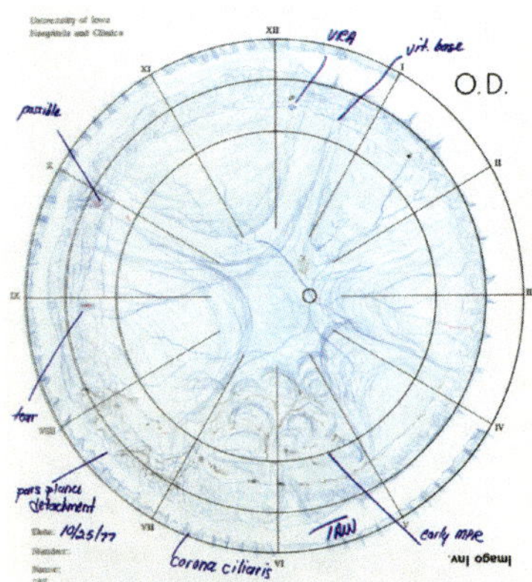

A.
From Simple to Complex Shading

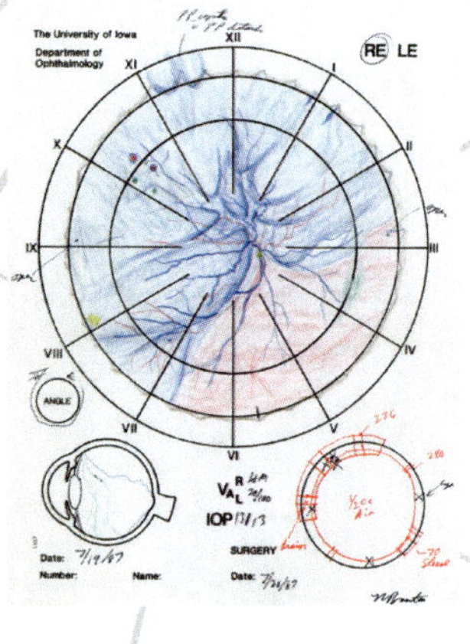

B.
Artistic Discipline and Combination Shading

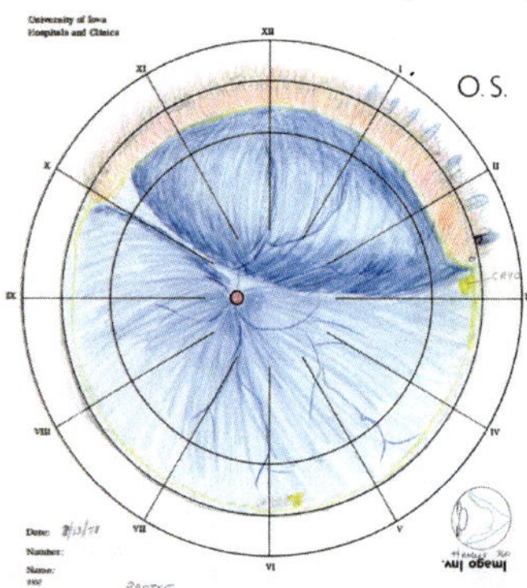

C.
Working the Pencil Stylistically: From One to Many Directions

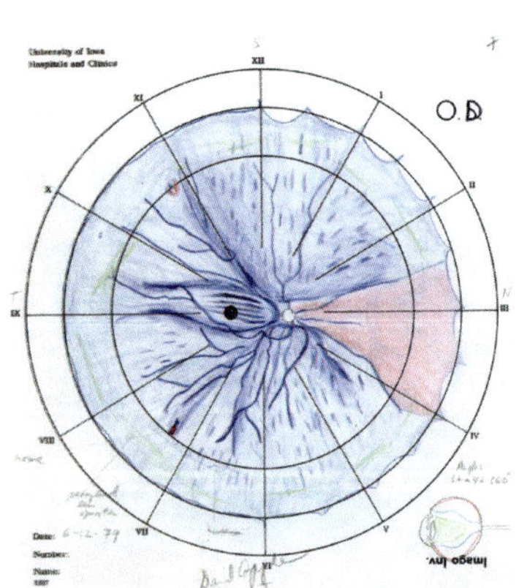

D.
Illustrating "Water" Under the Retina: A Maritime Motif

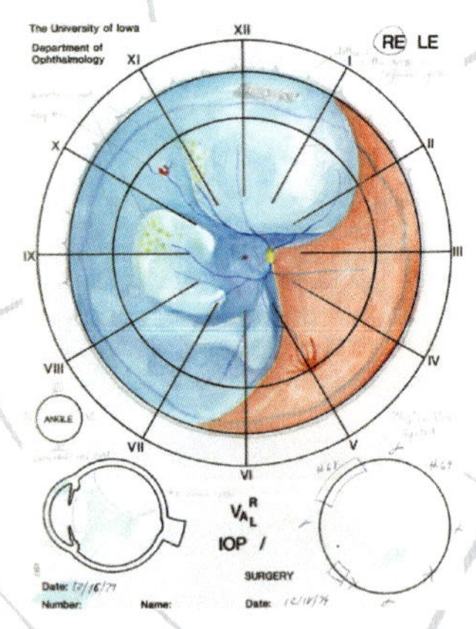

E.
Intentionally Artistic Use of Tone

Chapter 3

Gradations of Shading

A. From Simple to Complex Shading

Among the artists, various methods and levels of effort were taken in shading retinal detachments, shown as blue regions according to the standardized color code. Among the simplistic approach, illustrated here by Dr. Weingest, is the use of a wider principal outline in regions that appear darker. This simple attempt to represent the three-dimensional shape of the retinal detachment is not very visually effective but easy and quick to perform. Additional shading within the elevated regions of the retina does not precisely correspond to the clinical shading or to any conventional artistic convention.

"Almost inoperable," noted Tom Weingeist, MD, PhD (11/10/74)

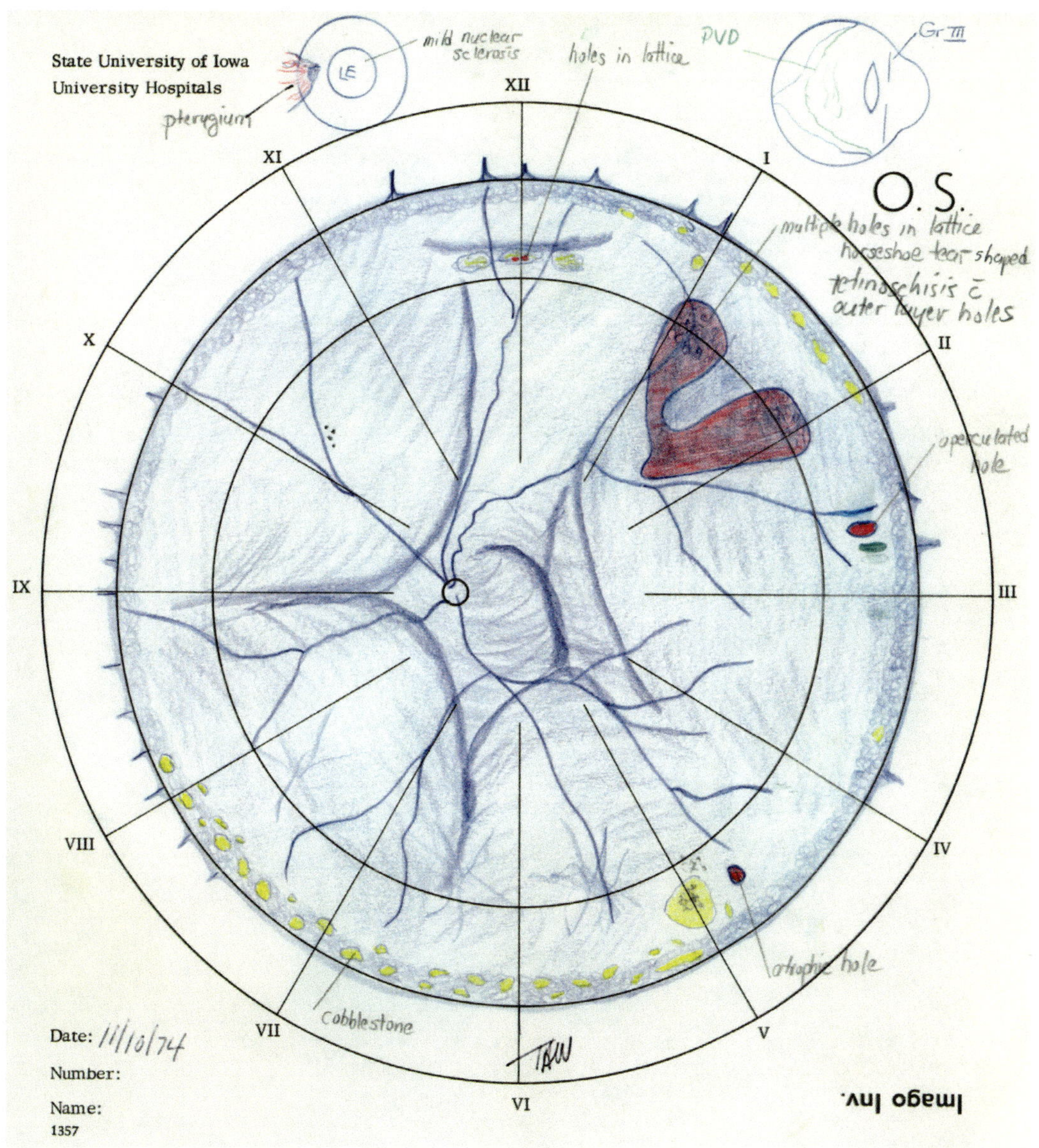

Artist Thomas A. Weingeist
November 10, 1974

Diagnosis:
Total retinal detachment, left eye, and retinoschisis with additional retinal holes [and] many star folds.

An extension of the principal line weighting is shown in this July 1987 image, where heavily weighted principal lines are broadened, representing a style midway toward the regional shading shown later in this chapter.

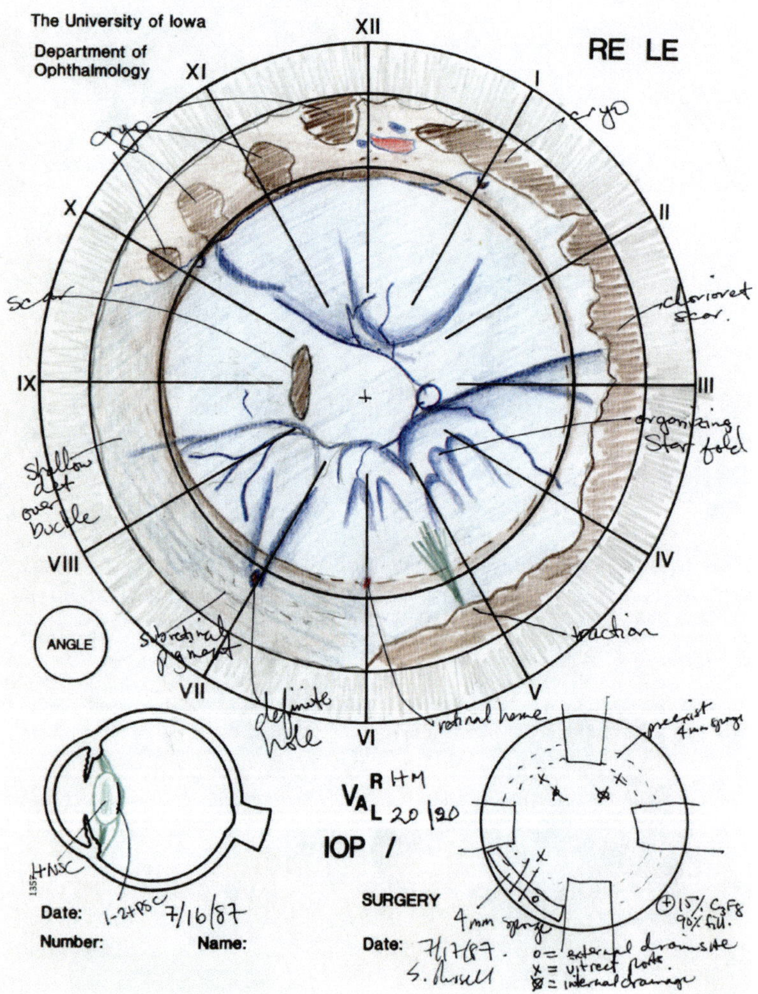

Artist Stephen R. Russell
July 16, 1987

Diagnosis:
Recurrent rhegmatogenous retinal detachment, right eye; hit in the right eye with a corncob fifty years earlier.

This drawing shows a more disciplined, iconic use of principal line shading. The retinal folds have a shading uniformity resembling evenly illuminated boulders.

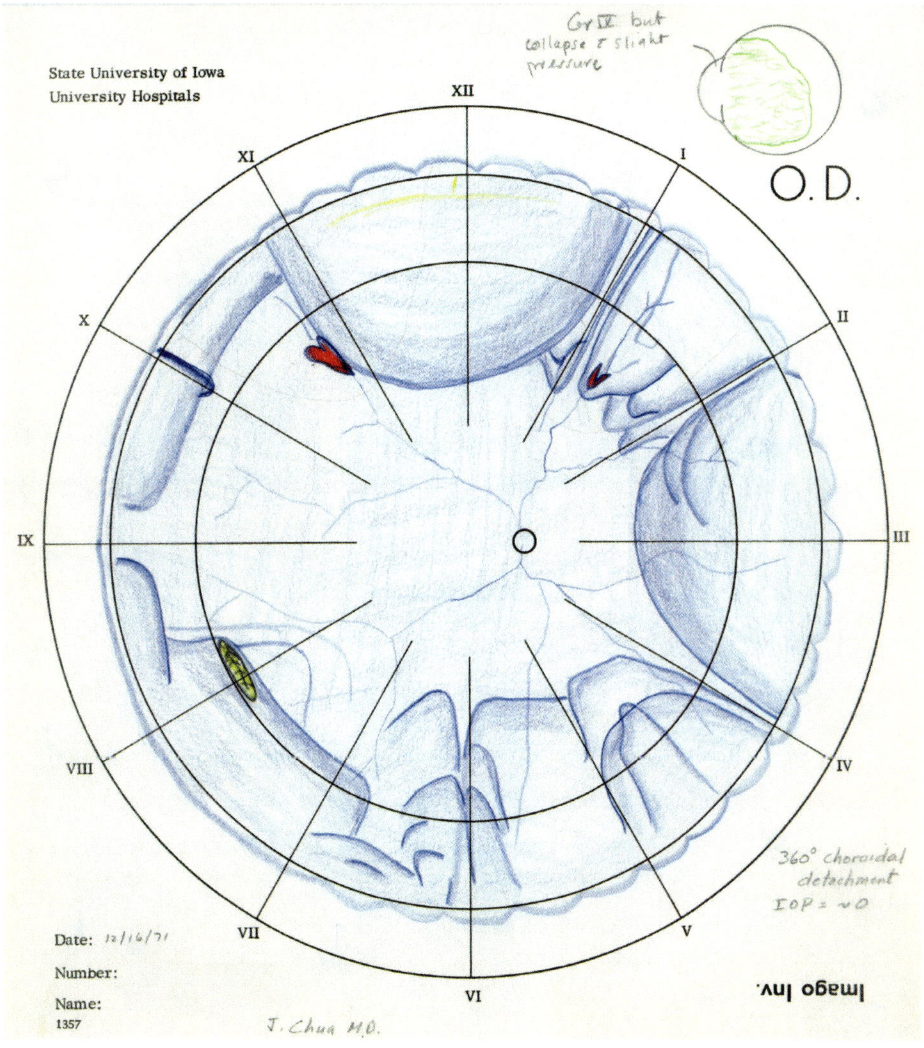

Artist Jonathan E. Chua
December 16, 1971

Diagnosis:
Total retinal detachment, right eye, four months following cataract extraction.

More elaboration of shading, as illustrated in this drawing from 1977, shows accessory lines of blue originating from the principal outlines that decrease in weight further from the principal line. The number of accessory lines and their curvature convey the contour and shading of these convex regions. Of note are the interesting contours illustrated at the bottom of this image, which could represent rarely seen retinal cysts (see the section called "Cysts" in chapter four).

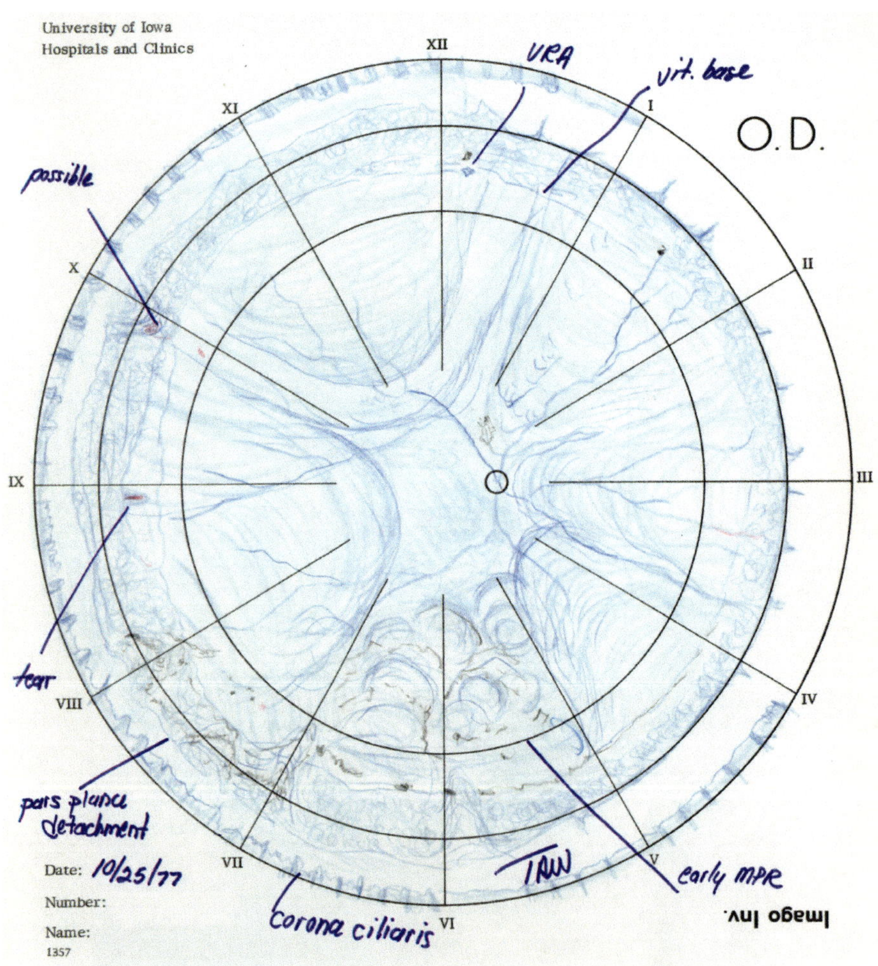

Artist Thomas A. Weingeist
October 25, 1977

Diagnosis:
Total rhegmatogenous retinal detachment, right eye. Early massive preretinal membrane formation; struck in right eye fifty years earlier.

Shading styles vary from earnest to frivolous. Instead of smooth variations in line weight, density, or number, in this image the artist uses a more primitive pencil technique: zigzag lines of dense blue. This effect is not an attempt to depict the abnormal retinal shape realistically, but it effectively conveys a sense of the tissue brightness.

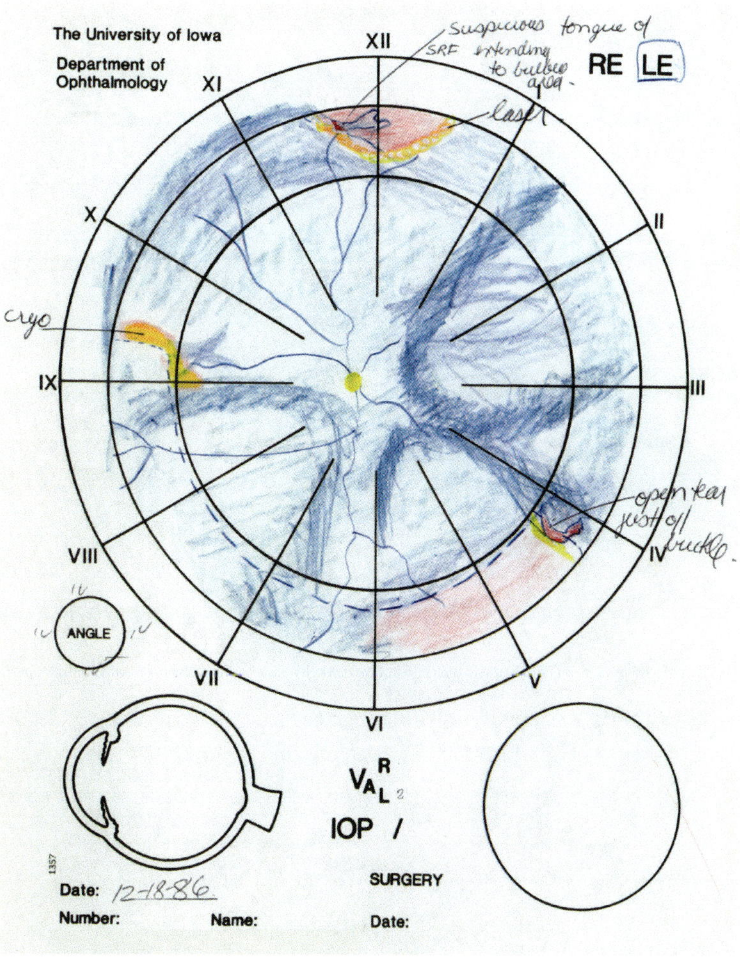

Artist E. Gary Servais
December 18, 1986

Diagnosis:
Recurrent retinal detachment, left eye, with previous procedures (retinopexy, pars plana vitrectomy, and scleral buckling with sponge), all in 1986, and previous pathology (proliferative vitreoretinopathy, choroidal effusion, and cataracts).

Just as some children apply greater force to their crayon or pencil to intensify its color, this artist used a heavy hand in shading. Here the shading styles are combined within regions to convey depth (or in the case of retinal detachment, height). However, the contrast range is narrowed when using heavy shading so the elevation of the detachment is less pronounced than with techniques shown in subsequent sections.

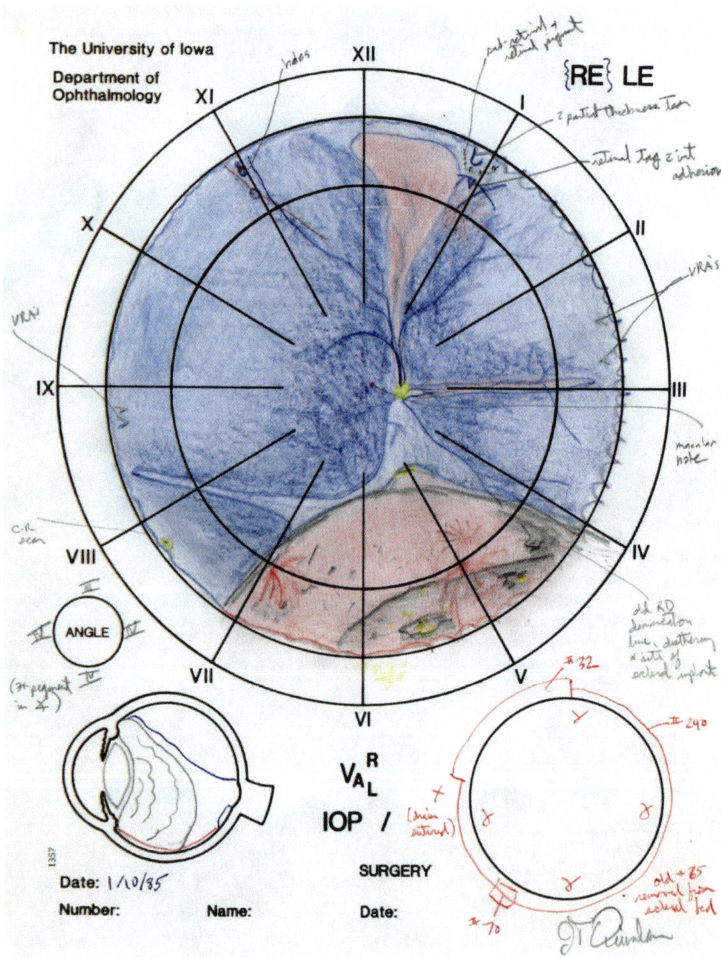

Artist James T. Quinlan
January 10, 1985

Diagnosis:
Recurrent rhegmatogenous retinal detachment, right eye.

B. Artistic Discipline and Combination Shading

Shown in the following four panels are the artists' rendering of combination shading, using principal line weight, accessory lines, junctional deviation of retinal vessel direction, and regions of denser color to convey retinal contour. Mixtures of these techniques are rarely found among regions within the same image. In nearly all drawings, each ophthalmologist—individuals with known tendencies toward obsessive-compulsive behavior—has not used more than one style of shading within each drawing.

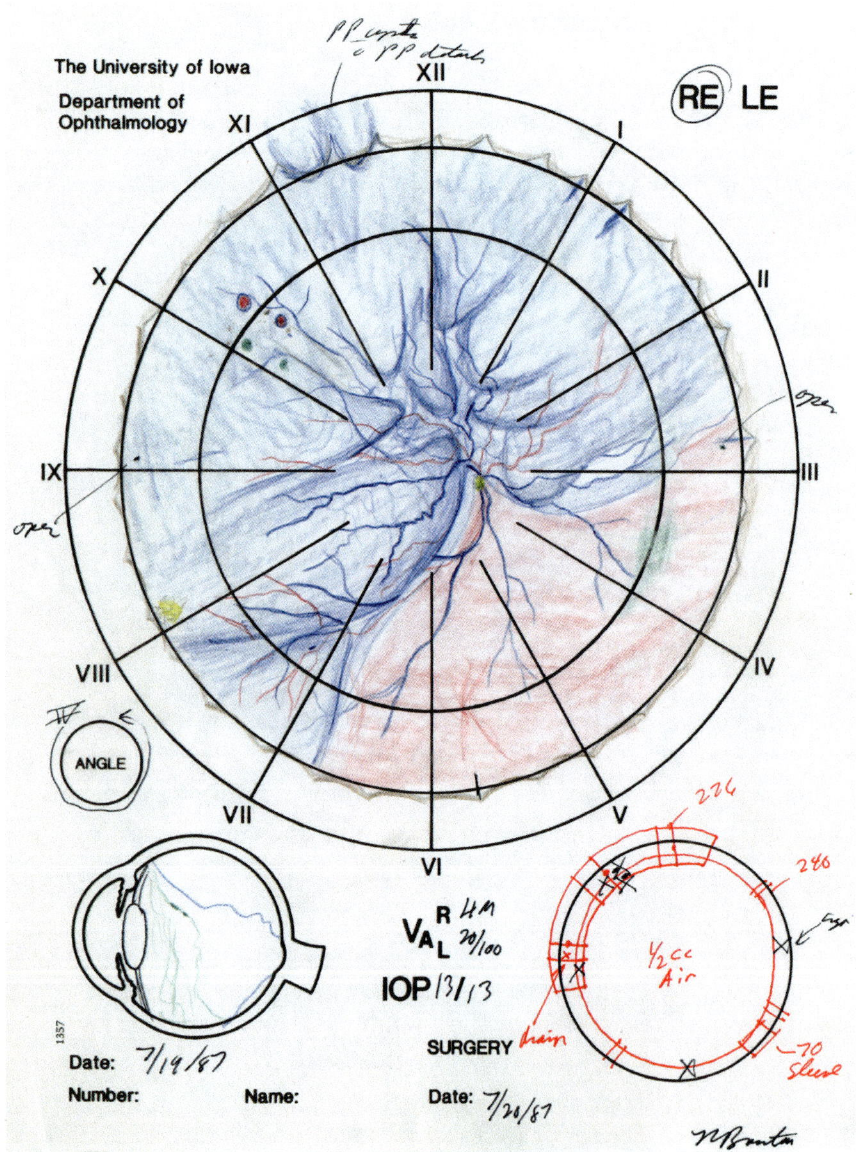

Artist Randall S. Brenton
July 19, 1987

Diagnosis:
Retinal detachment, right eye

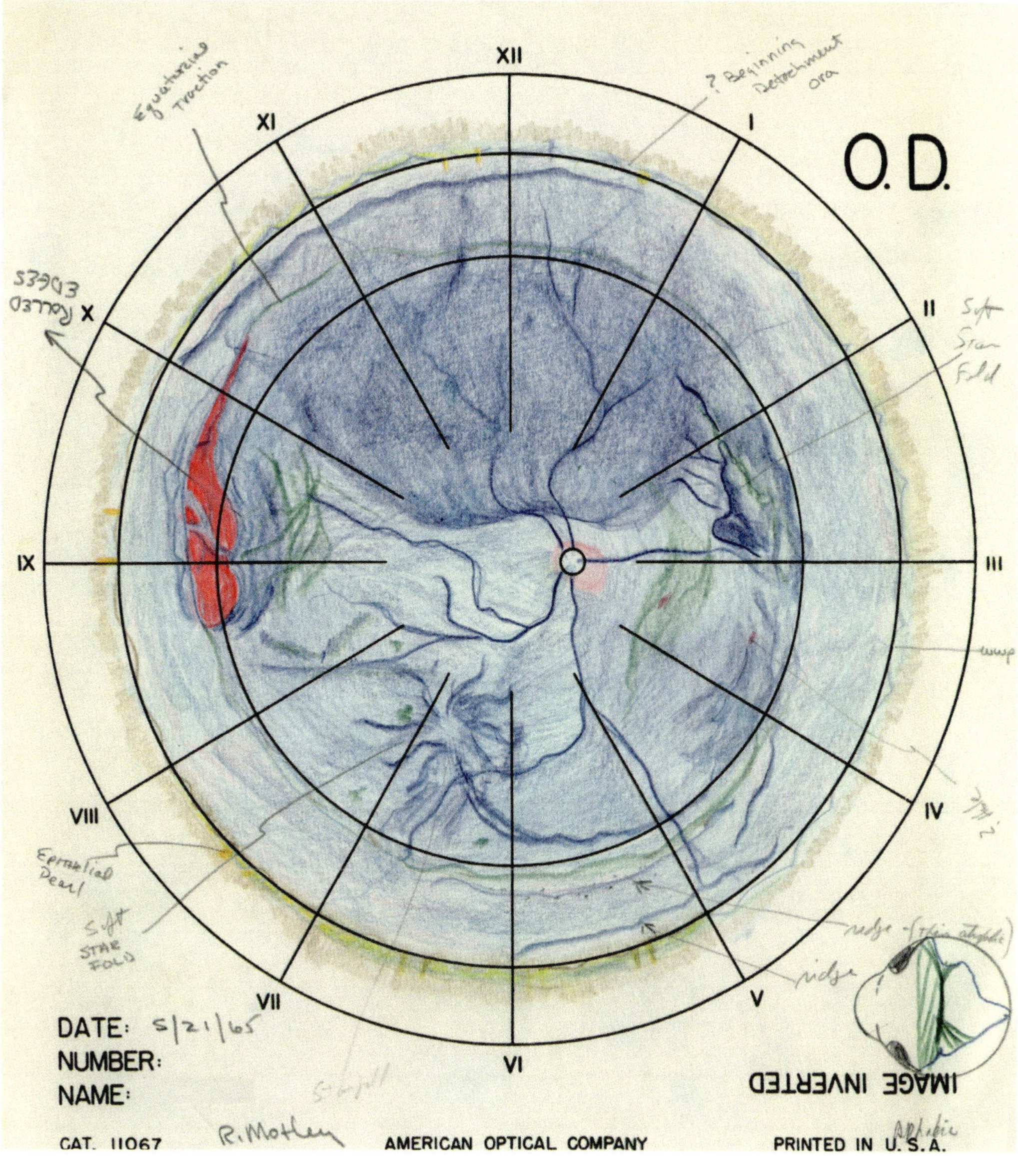

Artist Reid E. Motley
May 21, 1965

Diagnosis:
Total retinal detachment with large tear, right eye; previous cataract

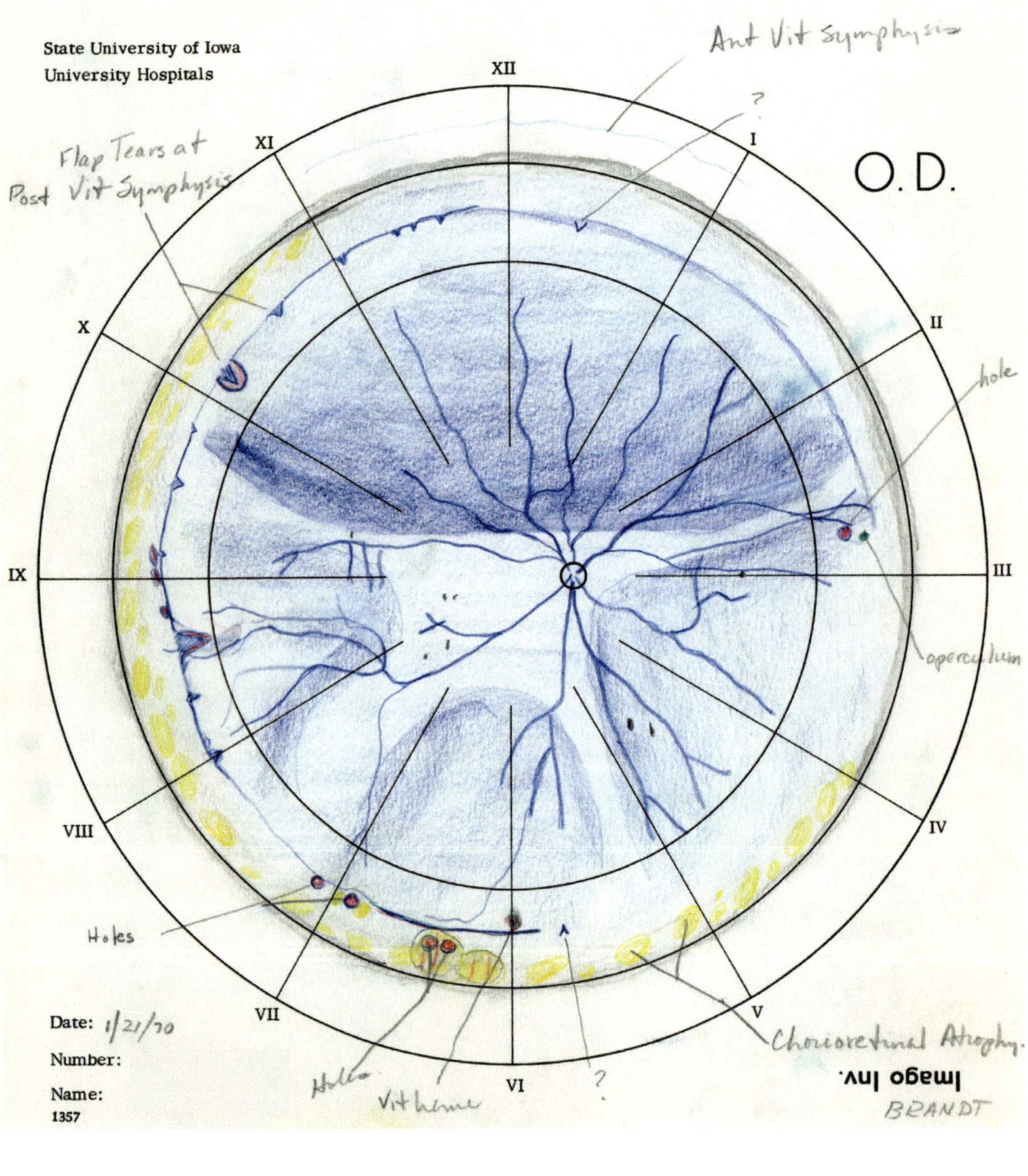

State University of Iowa
University Hospitals

O.D.

Date: 1/21/70
Number:
Name:
1357

Artist David E. Brandt
January 21, 1970

Diagnosis:
Retinal detachment with numerous round holes and flap tears, right eye, following cataract extraction and complete removal of the iris.

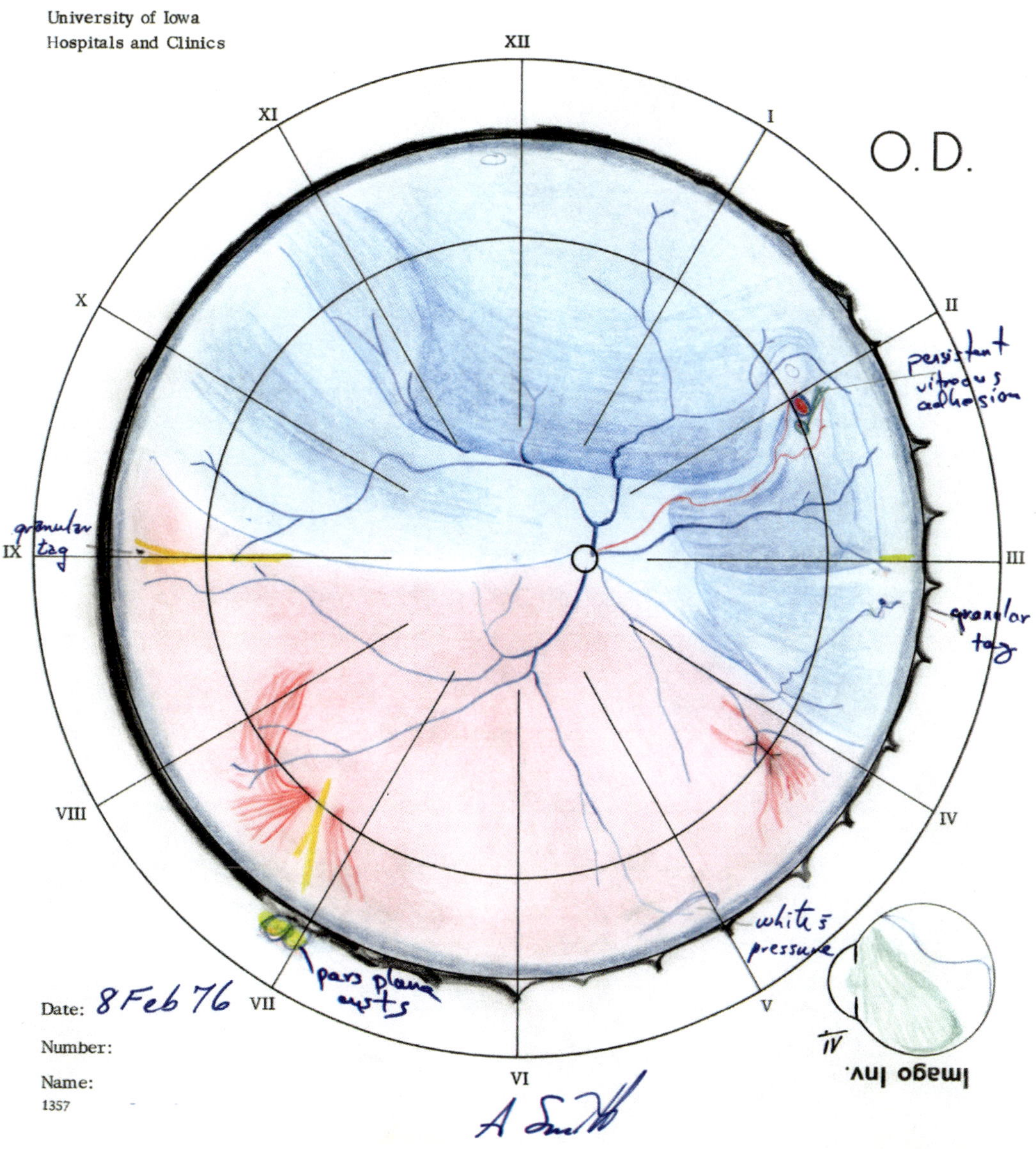

Artist Alexander G. Smith
February 8, 1976

Diagnosis:
Operculated hole (tear with retinal flap torn off) with secondary retinal detachment, right eye, following cataract removal nine years earlier.

C. Working the Pencil Stylistically: From One to Many Directions

In addition to blocks of color within regions, crossed or parallel lines were used to reinforce the perception of depth. In this November 1987 drawing, a predominantly radial line direction is used.

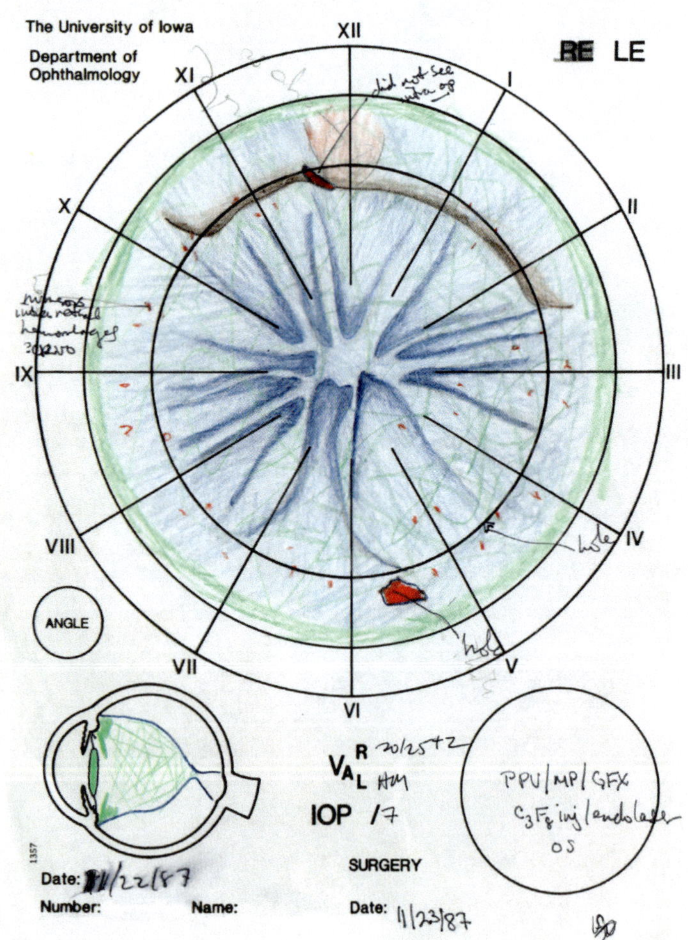

Artist Stephen R. Russell
November 22, 1987

Diagnosis:
Grade D3 proliferative vitreoretinopathy and close funnel total retinal detachment, left eye, with a history of pars plana vitrectomy.

In contrast to the previous image, this drawing includes individual folds of retina shaded with parallel lines as cylindrical structures. The orientation of cylindrical lines in multiple folds extends the visual effect circumferentially around the region of detachment in a similar fashion as in the previous image.

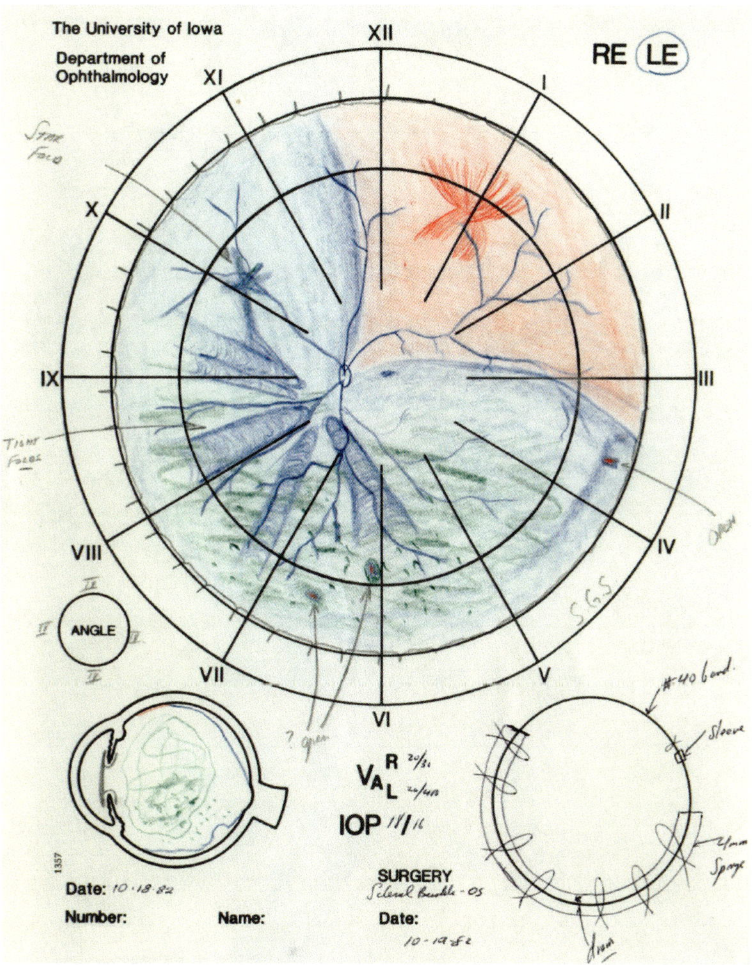

Artist Stanley G. Shortt
October 18, 1982

Diagnosis:
Retinal detachment, left eye.

Why not plaid shading? Here secondary lines enhance the principal outlines, but the midground and background are shaded in plaid. Variety seems to prosper in the retina clinic.

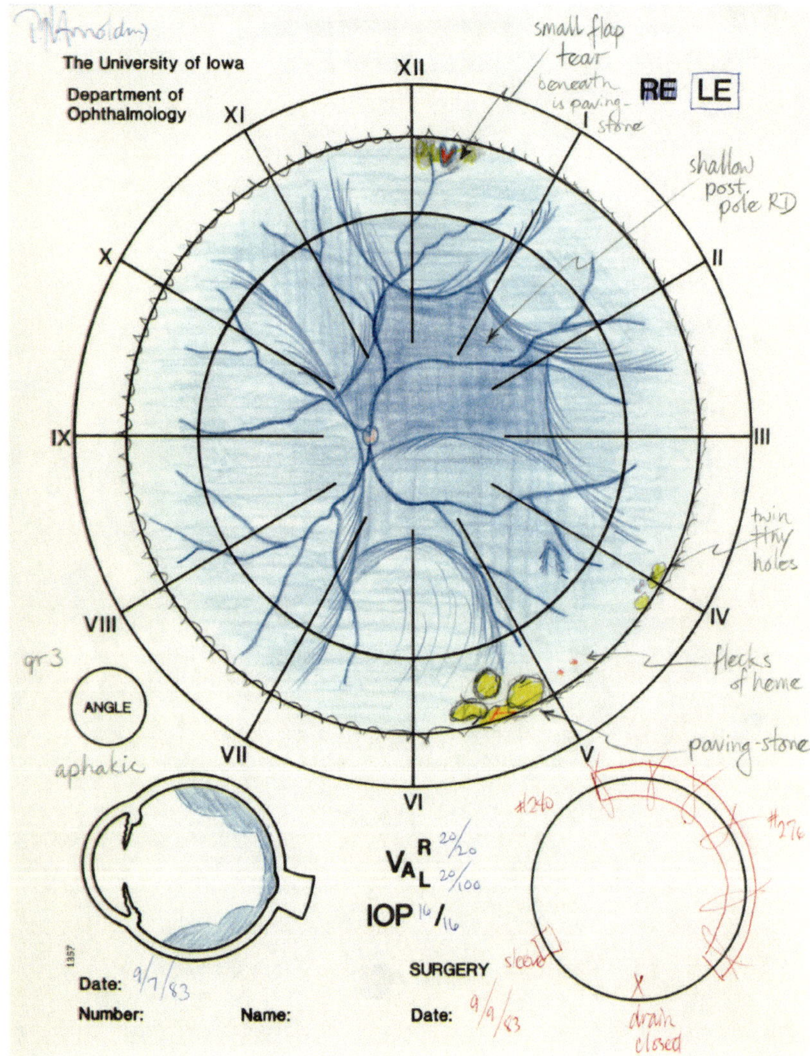

Artist Paul N. Arnold
September 7, 1983

Diagnosis:
Total retinal detachment with paving stone degeneration, left eye; cataract extraction two years earlier.

In this richly elaborated shading, the details of the peripheral retina are sparse, except in the superior-temporal quadrant. The retinal "tension lines," which are not actually seen, strongly suggest the bullous elevation of the superior and inferior retina. The pencil has been carefully worked to vary the weight of these curvilinear lines, with lighter weight over the apex of the detachment.

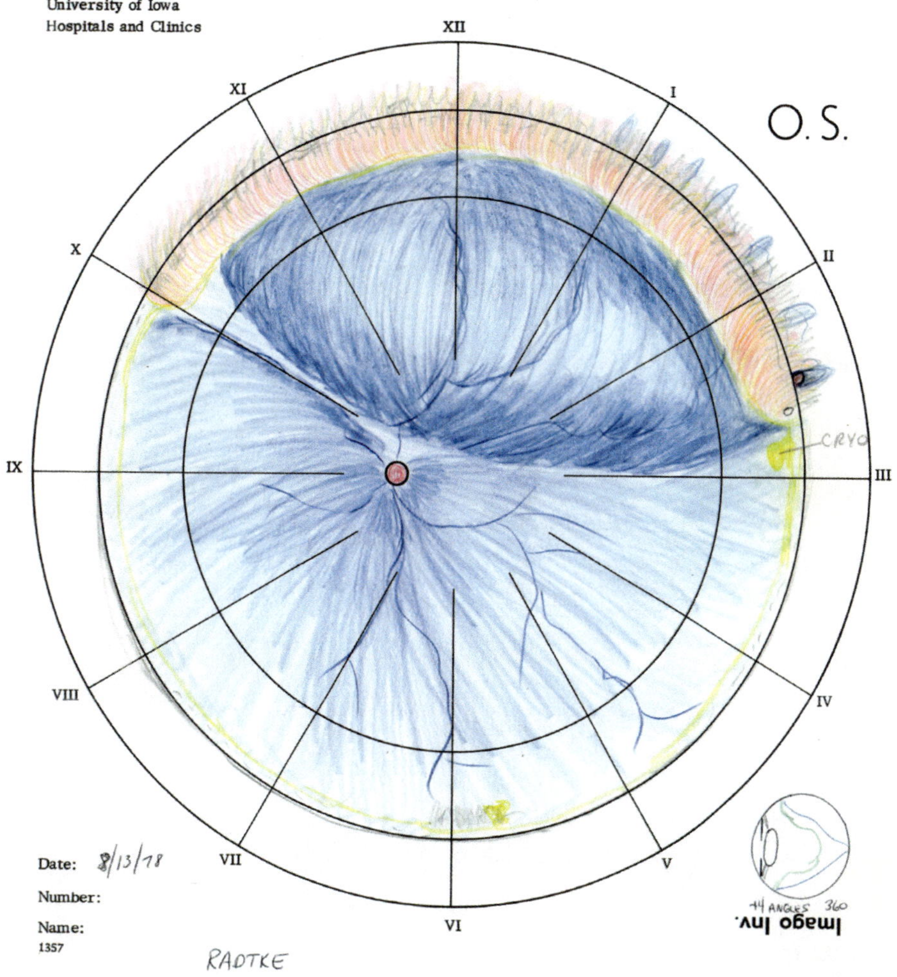

Artist Norman D. Radtke
August 13, 1978

Diagnosis:
270-degree recurrent retinal detachment, left eye, one month following scleral buckling.

"The degree of fatigue at the end of the day influenced a lot of these things."

Richard Dreyer, MD (3/1/11)

Texture can be added through color or through surface modification of the paper. In this drawing, the artist scratched curvilinear imprints into the paper that reinforced the blue lines interwoven with them. He also contrasted a radial shading of the attached retina (red) with oblique shading of the detached retina (blue).

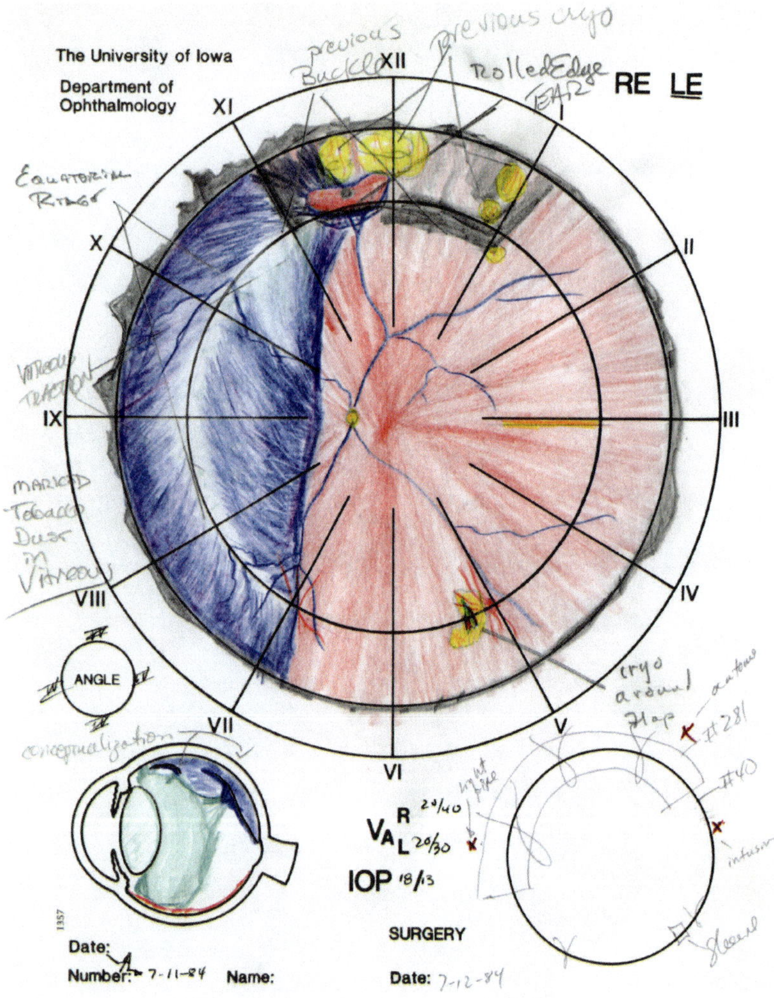

Artist Richard F. Dreyer
July 11, 1984

Diagnosis:
Recurrent retinal detachment, left eye, and proliferative vitreoretinopathy.

More subtle shading in two or more directions is shown in this image from 1978. Here the foreground appears to have been lightened by erasing. The textures are more delicate and the coloration range is enhanced through the overlay of varying hues of pencil color.

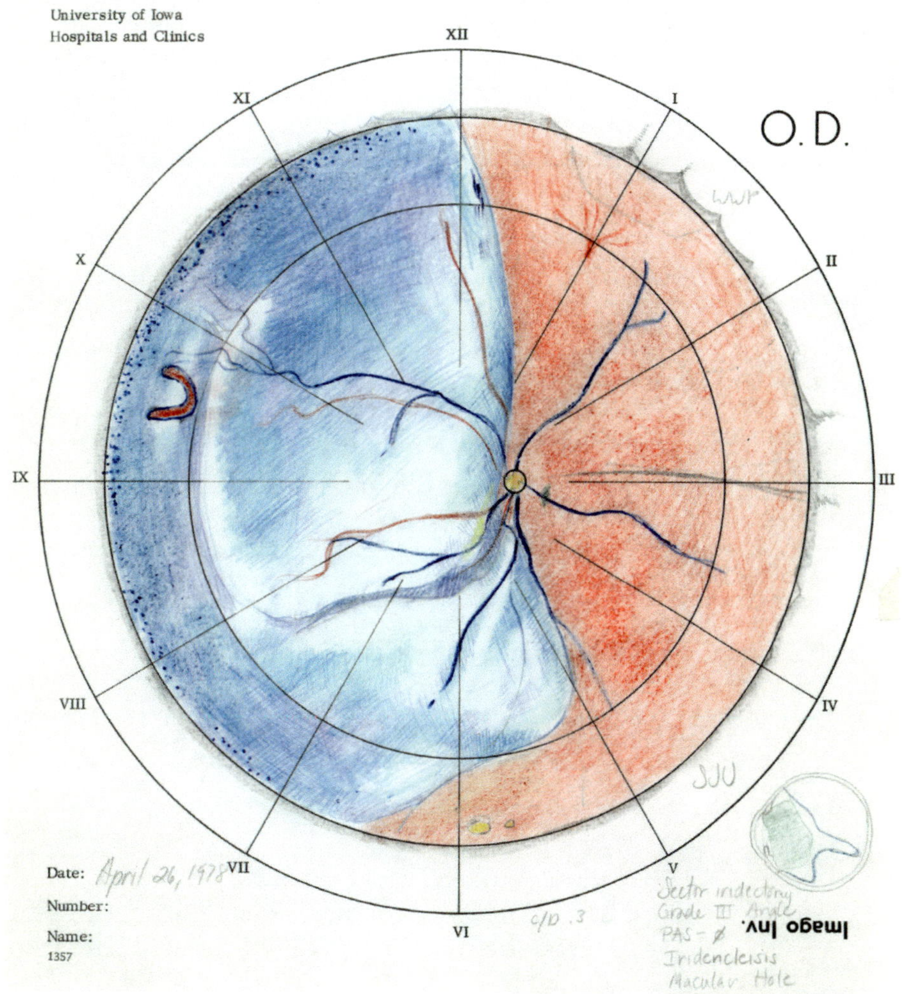

Artist Steven J. Vermillion
April 26, 1978

Diagnosis:
Retinal detachment, right eye, two months after cataract removal.

A drawing took "a couple hours, especially if you were compulsive."

Richard Dreyer, MD (3/1/11)

D. Illustrating "Water" Under the Retina: A Maritime Motif

One of the peculiarities of the color code may have to do with why blue was chosen to represent a detached retina. As a symbolic color, we associate blue with the ocean, lakes, streams, and other bodies of water. The subretinal fluid present under a retinal detachment is a watery accumulation. What may not seem obvious is why an artist would choose to use enhancing lines to suggest waves, undulations, or other dynamic characteristics associated with open bodies of water. In a progression of shading, this 1981 image shows wave-like shading of radial retinal folds.

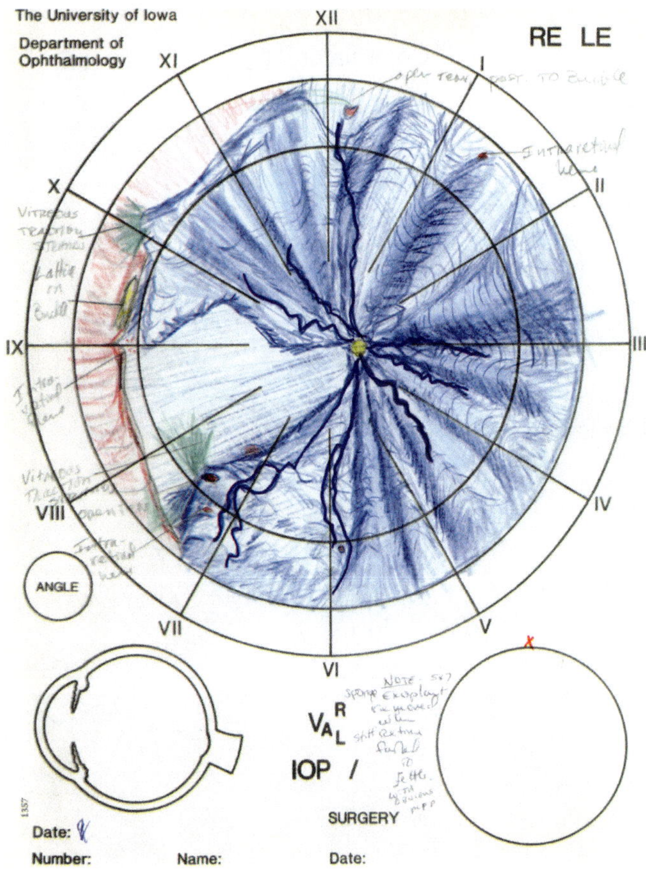

Artist Richard F. Dreyer
September 9, 1981

Diagnosis:
"Very difficult" three-quadrant recurrent retinal detachment with a flap tear superiorly and a small tear inferiorly, right eye.

"I used to love to do retina drawings like this one."

David Apple, MD (12/8/10)

In the progression of wave-like suggestion is the use of spaced, dashed lines, as in this drawing. A detached retina takes on a crenulated texture, thought to be due to hydration, which has the appearance of crinkled cellophane.

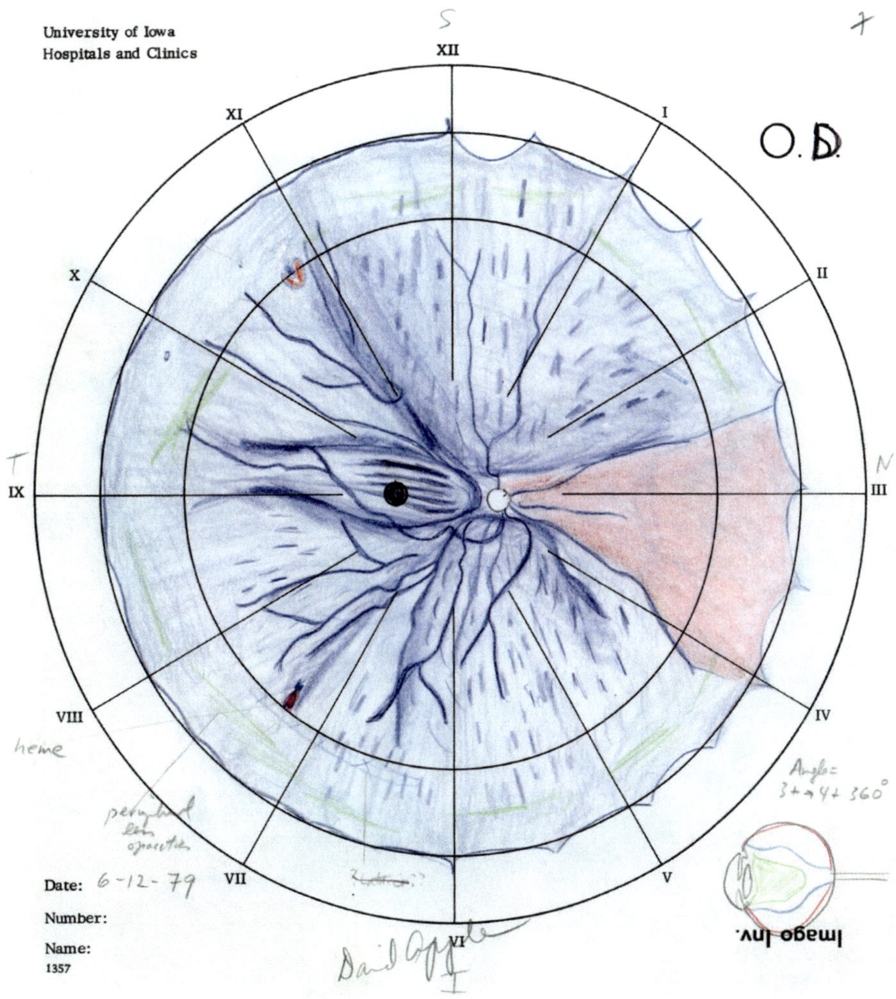

Artist David J. Apple
June 12, 1979

Diagnosis:
Rhegmatogenous retinal detachment, right eye.

"It is practically impossible to put such peripheral holes on a buckle."

William B. Snyder, MD (2/18/65)

Here, short, evenly spaced dashes may have been embellished as wiggles, much like waves. With eye movements, the fluid beneath the retinal detachment undergoes a to-and-fro movement resembling a wave or shimmer.

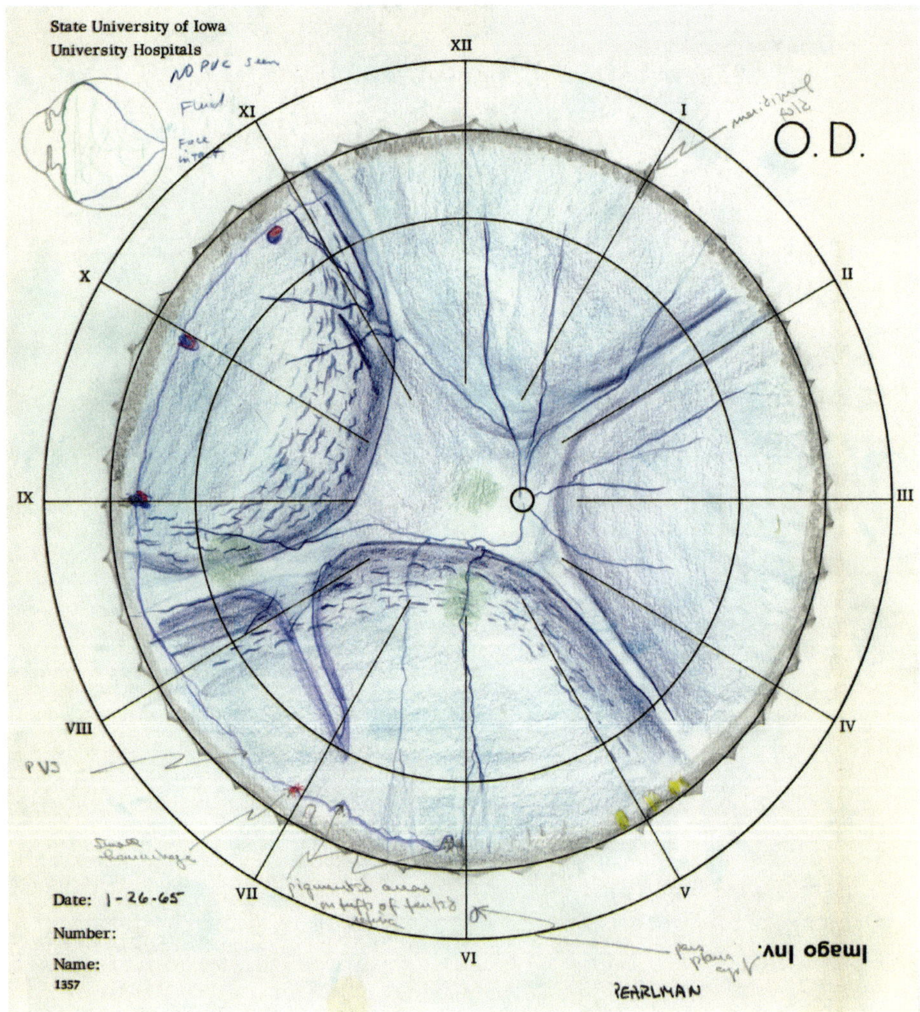

Artist Jerome T. Pearlman
January 26, 1965

Diagnosis:
Total retinal detachment, right eye, with "malignant appearing holes"; cataract extraction two months earlier.

"The patient was obviously very patient, but that's one of my favorite recollections of Iowans."

Francisco J. Pabalan, MD (12/6/10)

At some point the nautical character of the shading symbolism exceeds any direct resemblance to the retinal appearance. In this drawing, numerous waves are used.

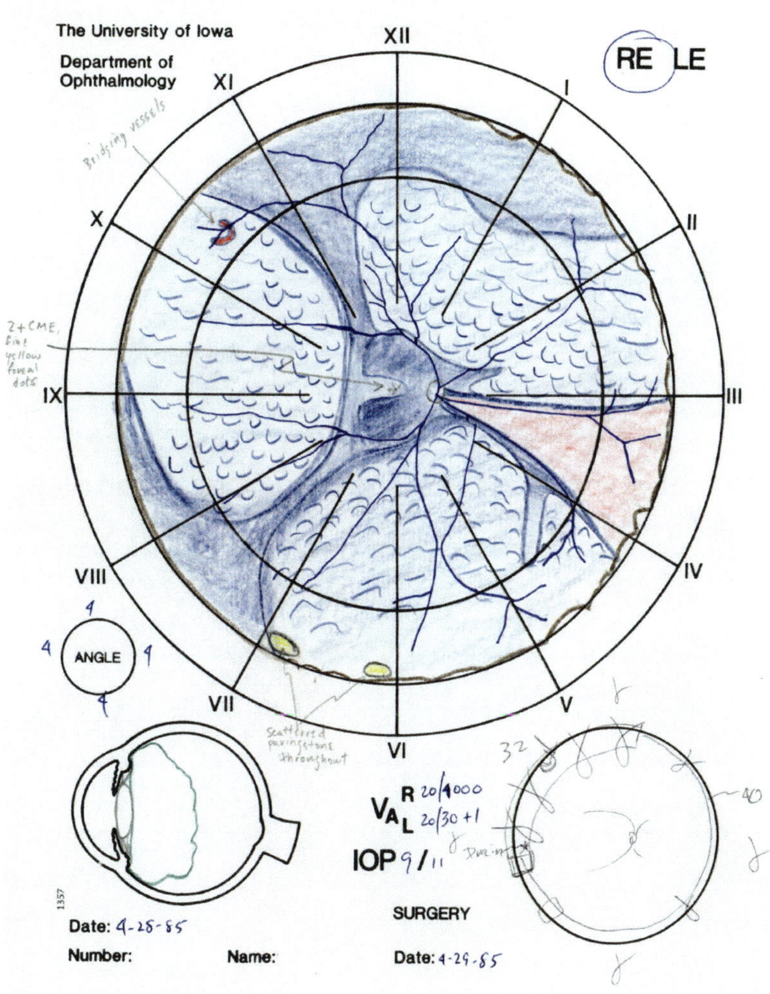

Artist Francisco J. Pabalan
April 28, 1985

Diagnosis:
Retinal detachment, right eye, following a fall on the ice three months earlier.

The nautical theme is extended with the use of reinforcing and refracting wavelets suggesting water movement. In this drawing, the enhanced diagrams of the vortex veins and ampullae (red objects) resemble multibranched invertebrates floating above and immersed in the sea.

"For a first-year resident entering the retina rotation, the choice of scleral depressor was important. We all wanted the best, as this would surely affect the detail of our drawings. There were thimble depressors, articulated depressors, double-ended flat depressors, etc. I was equally not at home with any of them, but remained convinced an overpriced name-brand depressor was vital to the exam. I sought advice about a detachment from Dr. Vern Hermson. Dr. Hermson had done retina surgery in private practice for years, returning as a fellow in order to have official certification. He reached into his pocket for a depressor, and not locating it, I offered mine. He politely declined, picked up a paper clip, and gently performed a 360-degree indentation. As it turned out, the depressor missing from his pocket was not made of stainless steel. He was looking for a Q-tip. On numerous subsequent occasions, I watched him roll the Q-tip around the eye in two uninterrupted, 180-degree arcs, completing a peripheral retina exam to the ora serrata in less than thirty seconds. My exam, to this day, consists of what seems to be 360 separate indentations."

Kent Stiverson, MD (12/16/10)

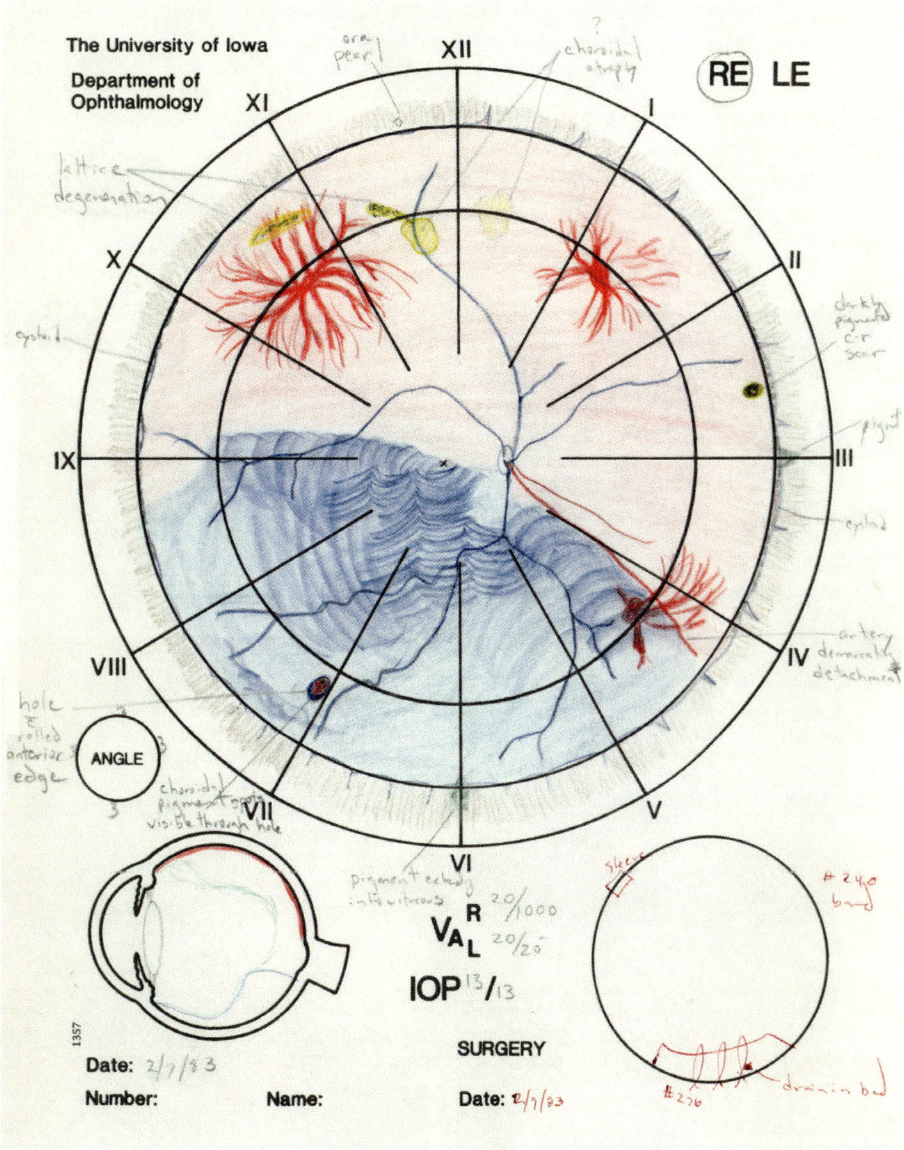

Artist R. Kent Stiverson
February 7, 1983

Diagnosis:
Rhegmatogenous retinal detachment, lattice degeneration, and two small holes, right eye.

E. Intentionally Artistic Use of Tone

Especially when ophthalmologists had experience or training in art prior to learning fundus drawing, as did those who created three of the four images in this section, color values can be varied to suggest foreground, middle ground, and background. In this image from December 10, 1979, the retinal drawing medium is watercolor with lighter tints used for the foreground and darker ones for the background or oblique surfaces. This becomes most evident when examining the areas of attached retina, or red portions of the image, in which the closest to the observer is nearly white. Similarly, over the detached retina, shown in blue, the portion in the foreground (highest elevation of retina) is represented in white. The color values do not directly correspond to the clinical appearance for either of these regions, but match the coloration that would be expected if following the artistic rules for perspective and shading.

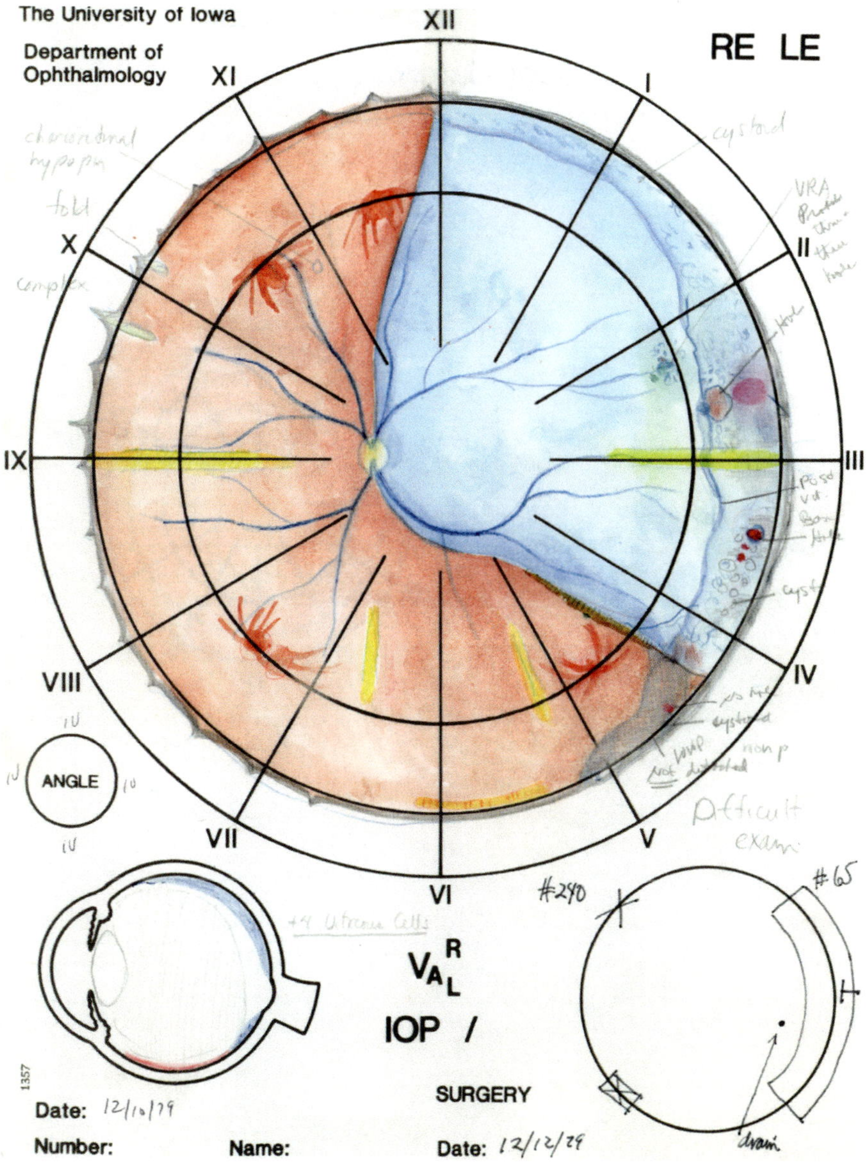

Artist Steven J. Vermillion
December 10, 1979

Diagnosis:
Rhegmatogenous retinal detachment, left eye.

In a more developed fundus drawing from the same artist, the shading corresponds more closely with the clinical hue values of the foreground, mid-ground, and background. Of note here is that the shading creates a sensation of surface contour that, despite its artificial coloration (of blue for the detached area), matches the clinical shading.

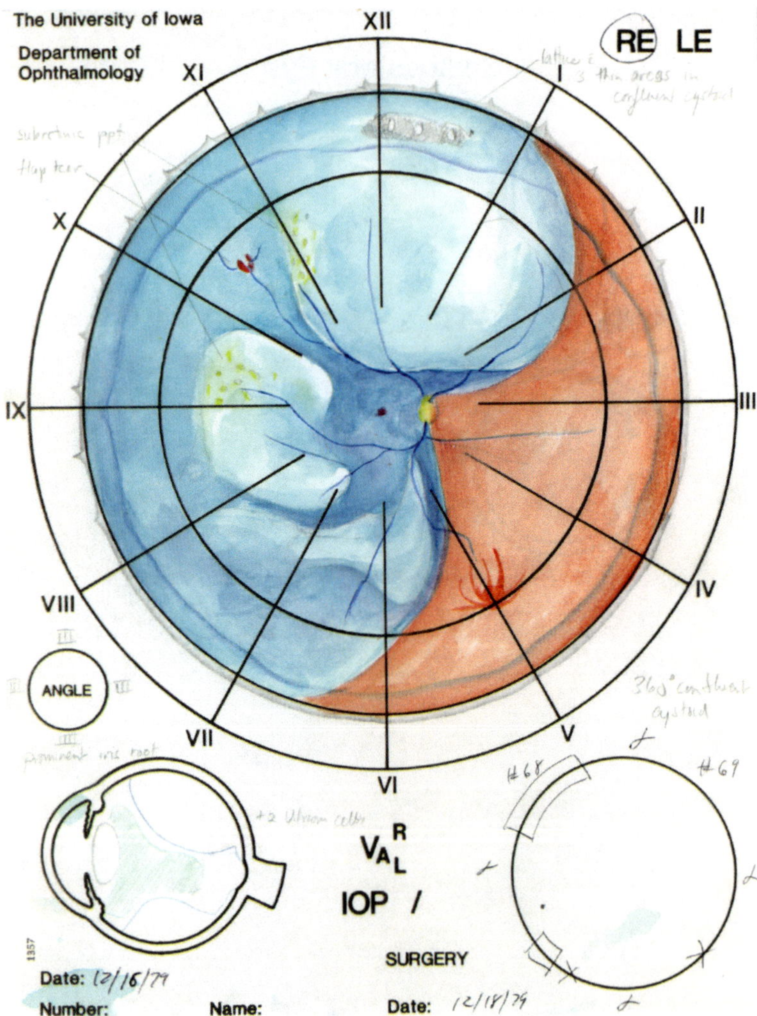

Artist Steven J. Vermillion
December 16, 1979

Diagnosis:
Rhegmatogenous retinal detachment, right eye.

Another approach taken by an artist who, prior to ophthalmology, had no formal experience or training as an artist is this example in which the subtle figuration of the retinal surface mimics the fine vascular congestion or retinal crenulations seen in detached retinas. The image, while figurative, has a measured and delicate detail.

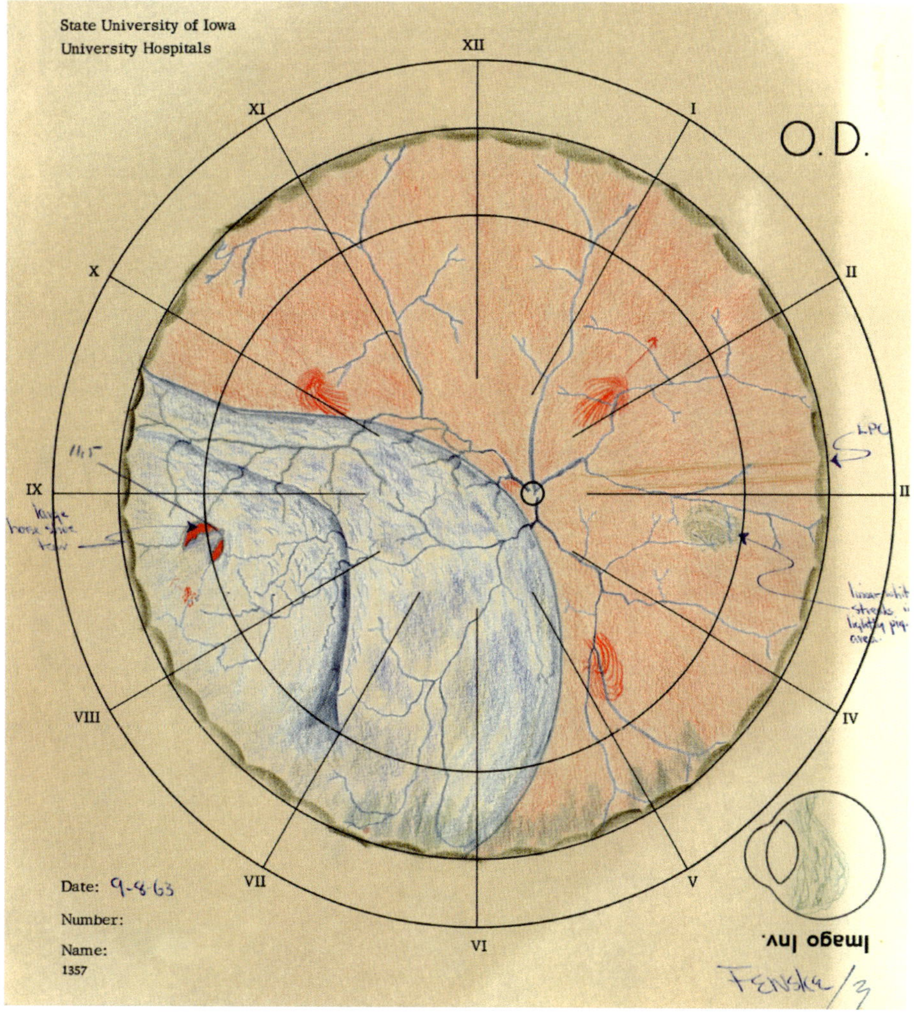

Artist David Fenske
September 8, 1963

Diagnosis:
Inferior temporal retinal detachment, right eye.

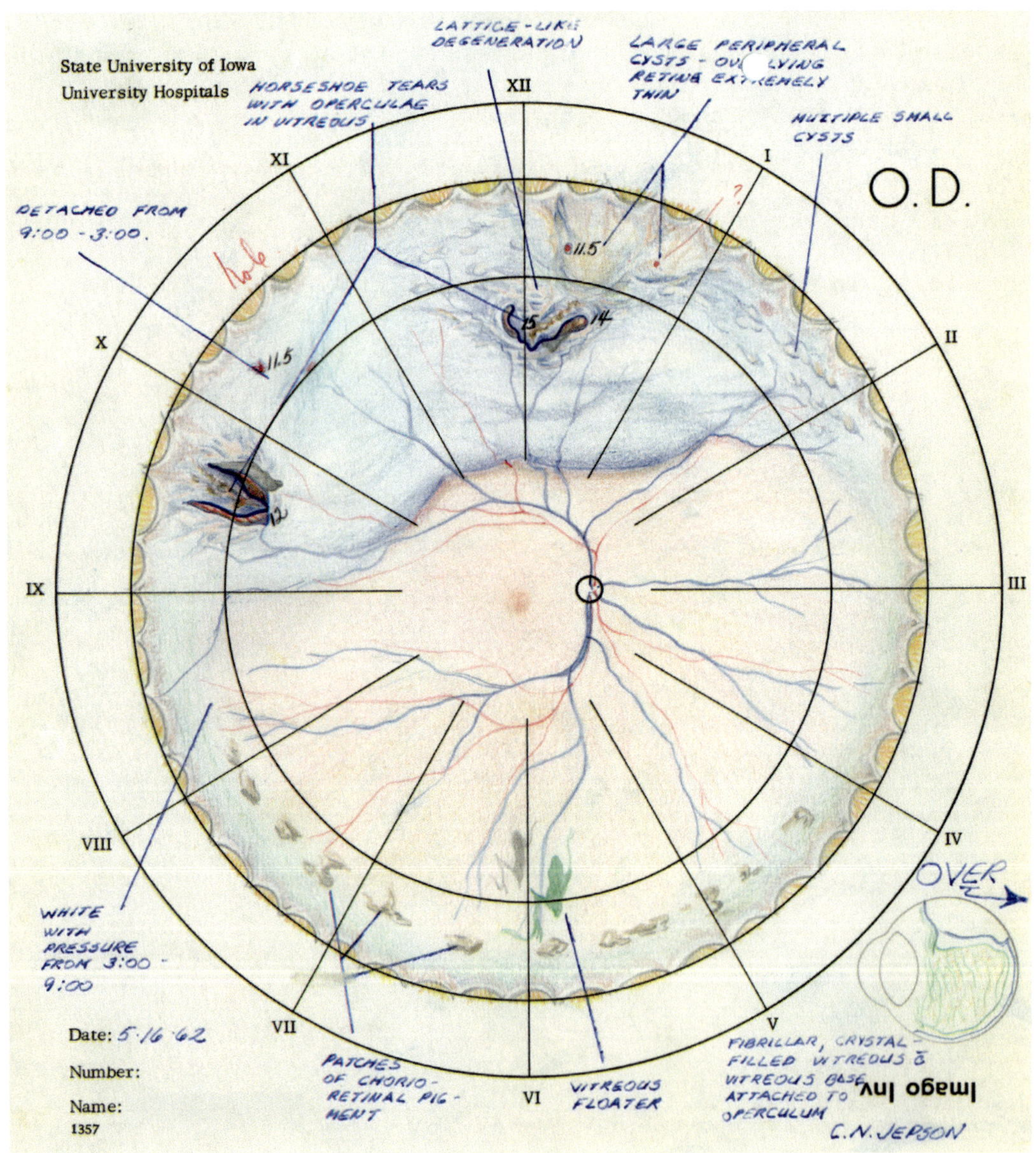

Artist C. Neal Jepson
May 16, 1962

Diagnosis:
Retinal detachment, right eye.

In the most fully developed form, the shading within the detachment has highly complex combinations of regional color and enhancing lines for emphasis. The care taken in detailing the ora serrata, even in the regions of attached retina, is noteworthy.

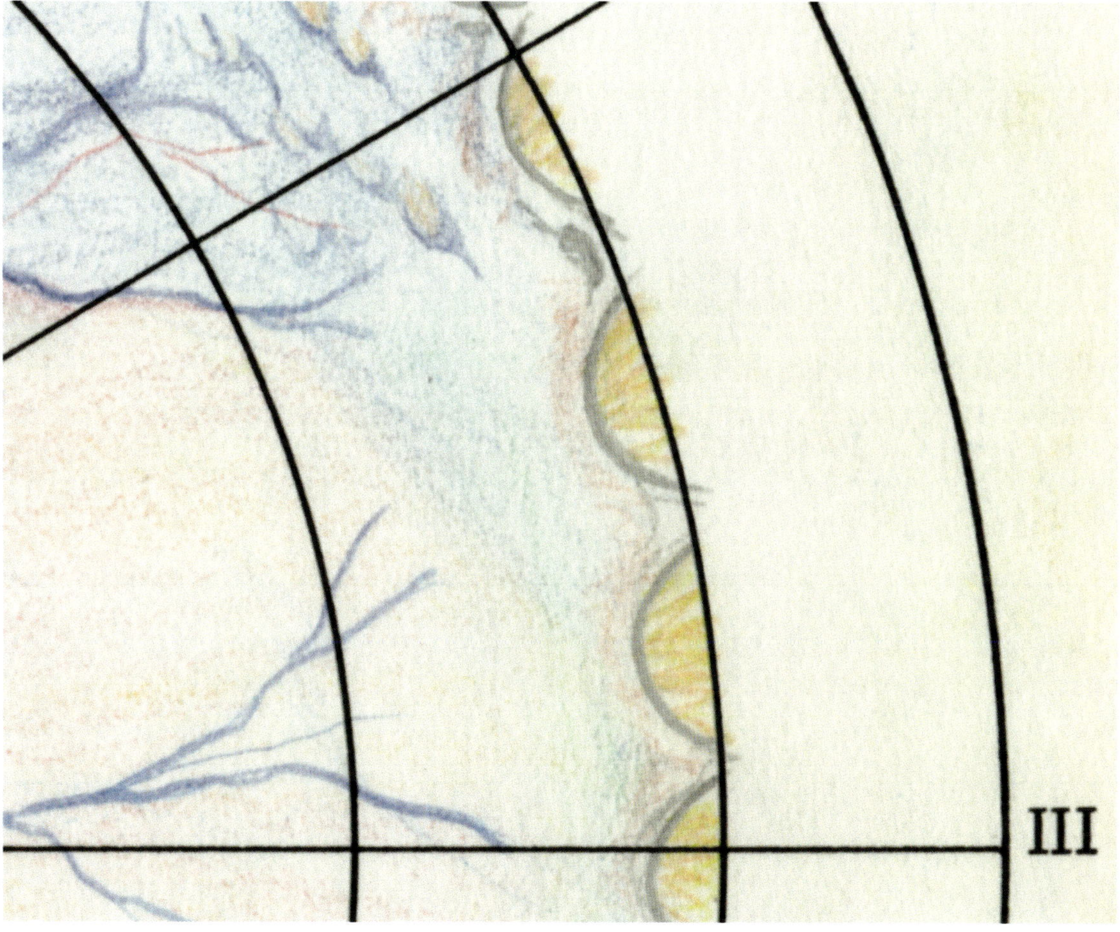

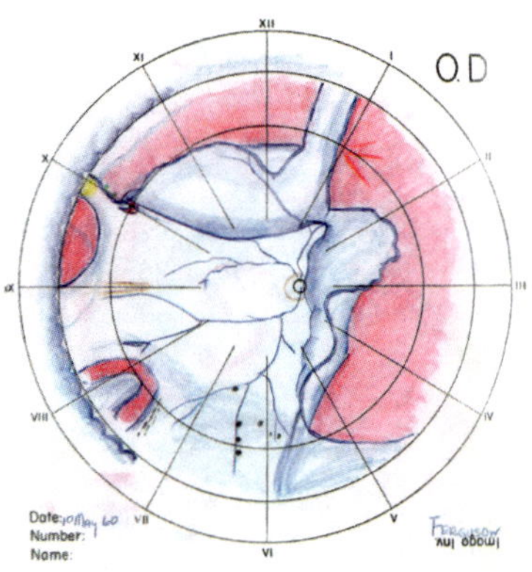

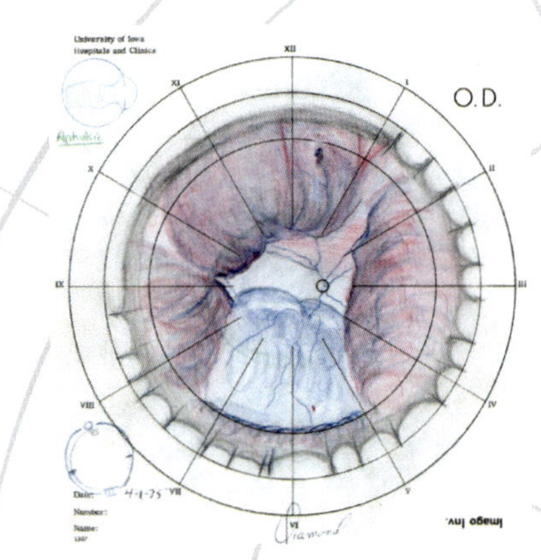

"Efficiency of a practically flawless kind may be reached naturally in the struggle for bread. But there is something beyond—a higher point, a subtle and unmistakable touch of love and pride beyond mere skill; almost an inspiration which gives to all work that finish which is *almost art—which is art."*

Joseph Conrad, The Mirror of the Sea (1906)

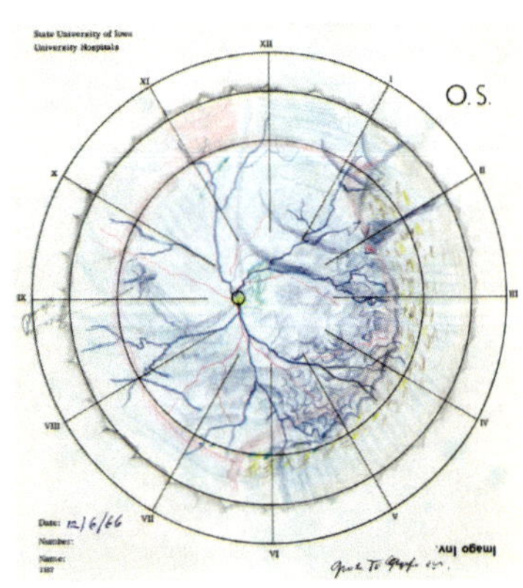

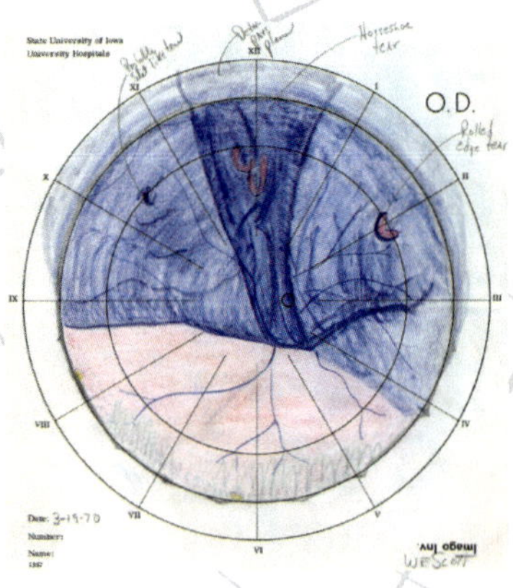

Chapter 4

The Representational Range of Fundus Abnormalities

"That finish," of which Conrad speaks in the epigraph on the previous page, is the completeness or maybe the metaphorical polish that sets the drawings in this book apart from the rest.

These drawings, like all fundus drawings, were done as part of the patients' medical care to record, diagnose, and treat pathological conditions of the retina and surrounding structures. The drawings were not done solely to express the artists' vision; however, they do express that vision, in medical and artistic terms. The artistic merit of these particular drawings led to their inclusion in this book, then, artistic merit that is captured beautifully in Conrad's words: "a higher point, a subtle and unmistakable touch of love and pride beyond mere skill...which is art."

More than any other place in *The Lost Art of Retinal Drawing*, including the chapter that explains the artists' process, this chapter reminds readers that the art pieces represented here began as medical illustrations. Although this fact is not concealed elsewhere in the book, it is also not highlighted, until now.

The drawings displayed on the following pages reveal how particular artists depicted their patients' retinal detachments, giant tears, cystoid degeneration, retinal cysts, and choroidal detachments, as well as their patients' proliferative vitreoretinopathy accompanied by star folds. Medically, the drawings reveal unique characteristics since each retina is different. Artistically, the drawings reveal stark stylistic differences since each artist created his or her representation of an individual retina in his or her own unique way.

As mentioned earlier, the creation of these stylish medical illustrations was before 1990, but since then the purpose has changed. Like a 1752 Vincennes Porcelain lidded bowl and dish (Getty Museum) and a seventh-century B.C. Etruscan terracotta vase (New York Metropolitan Museum of Art), the fundus drawings served a practical purpose and now can be celebrated for their artistry, for that "something beyond."

In the various sections that follow, certain anatomical features are accentuated and others are not. Some images include notes about the individual; others do not. Yet each section includes drawings depicting a diagnosis or feature. "Having a separate [section] for cystoid degeneration seems contrived," Dr. Watzke stated in an e-mail about this book (4/26/11). "It really is not a significant part of retinal anatomy at least as far as a retinal detachment is concerned. As you know, it is normally present but some examiners accentuated it in their drawings; some did not."

What Dr. Watzke said is both a true statement and a major reason that cystoid degeneration, in which cells on the periphery form cysts, remains a section in this chapter. Even artists who draw the ocular fundus have their own styles and include or emphasize certain features over others. The varied representations of, or disregard of, cystoid degeneration and other features, such as cobblestone degeneration, star folds, and spidery vortex ampullae indicate artistic intent, despite the clinical insignificance of some formations.

Giant Tears and Dialysis

Giant retinal tears are defined as tears that are three or more clock hours (one quadrant) in circumference. In some cases an examiner may have difficulty determining if the retina is torn from itself or from its junction with the ciliary body at the ora serrata. Separation of the retina at the ora is referred to as a retinal dialysis. In this image, the retina has been torn from the ora resulting in a giant retinal dialysis, a disorder that was extraordinarily difficult to fix at the time of this drawing and unlikely to have been successfully repaired.

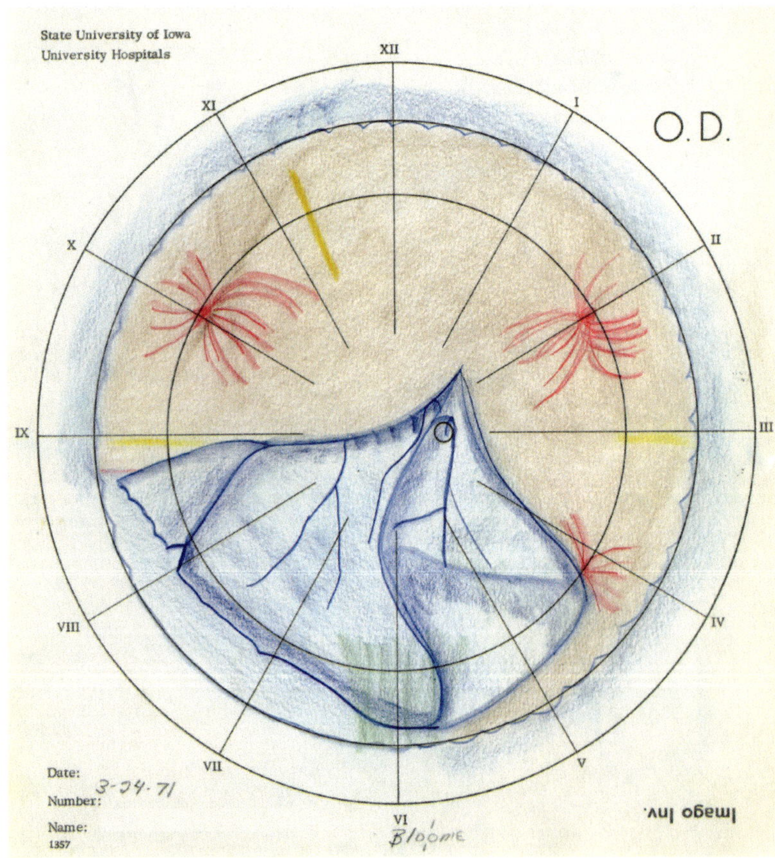

"This 270-degree giant tear wasn't fixable at that time, but forty years later we can reattach the majority of cases like this."

Mike Bloome, MD (12/2/10)

Artist Michael A. Bloome
March 24, 1971

Diagnosis:
"A severe form of retinal detachment known as a giant retinal tear," right eye (stated by Tom Burton in a letter to an insurance company that had claimed this detachment must be from trauma); cataract extraction six months earlier.

This giant retinal tear has a peculiar petal-like contour along its posterior or central edge. The peripheral edge is contrastingly jagged. Although these drawings may be beautiful, the experienced retinal surgeon is inwardly anxious examining these images, anticipating complications and failure. The nonsurgeon is not typically aware of this tension.

"In those days, the indirect ophthalmoscope was not as refined as today."

Harry Stephenson, MD (1/3/11)

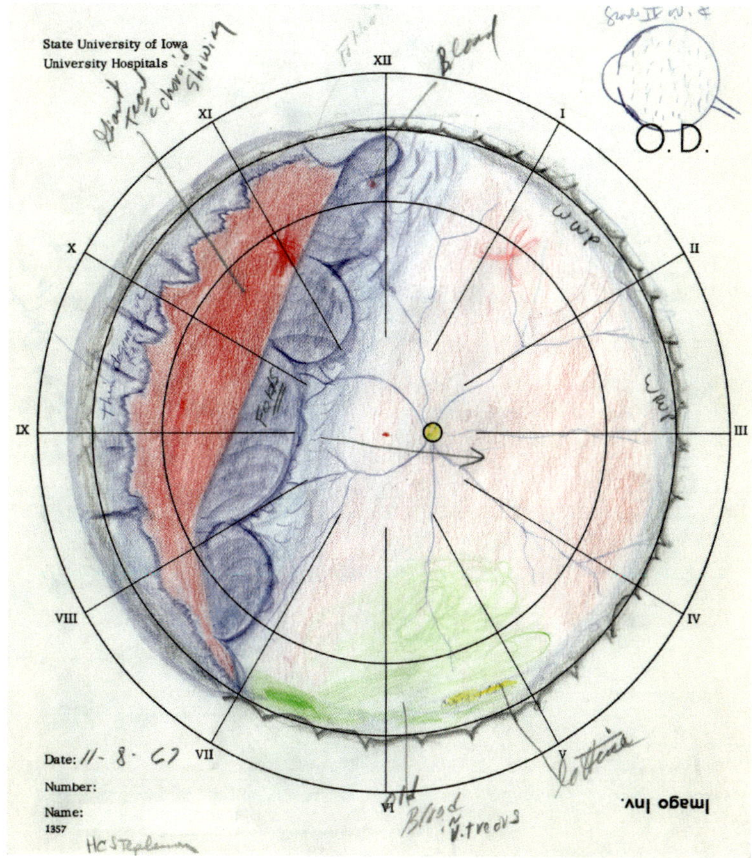

Artist Harry C. Stephenson
November 8, 1967

Diagnosis:
Giant retinal break, right eye, ten months after having a cataract extraction and retinal hemorrhage.

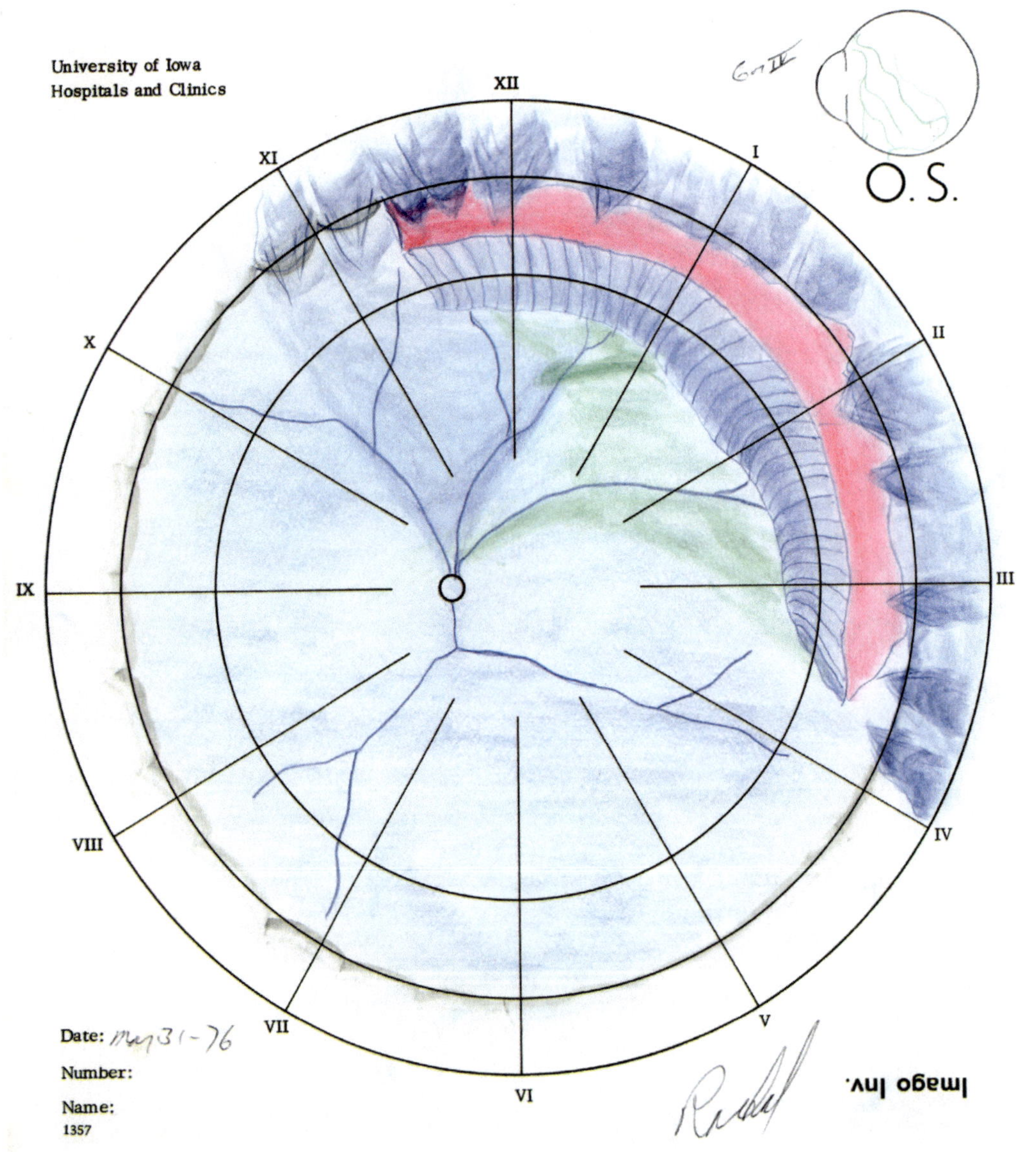

Artist William F. Rachal
March 31, 1976

Diagnosis:
Total retinal detachment with giant tear (twelve to three o'clock), left eye; cataract extraction nine weeks earlier.

This image illustrates, in finely detailed fashion, the rolling or curling that develops at the edge of many giant retinal tears. The artist also drew the outlines of the ciliary processes (the darker blue, arrow-like structures in front of the tear).

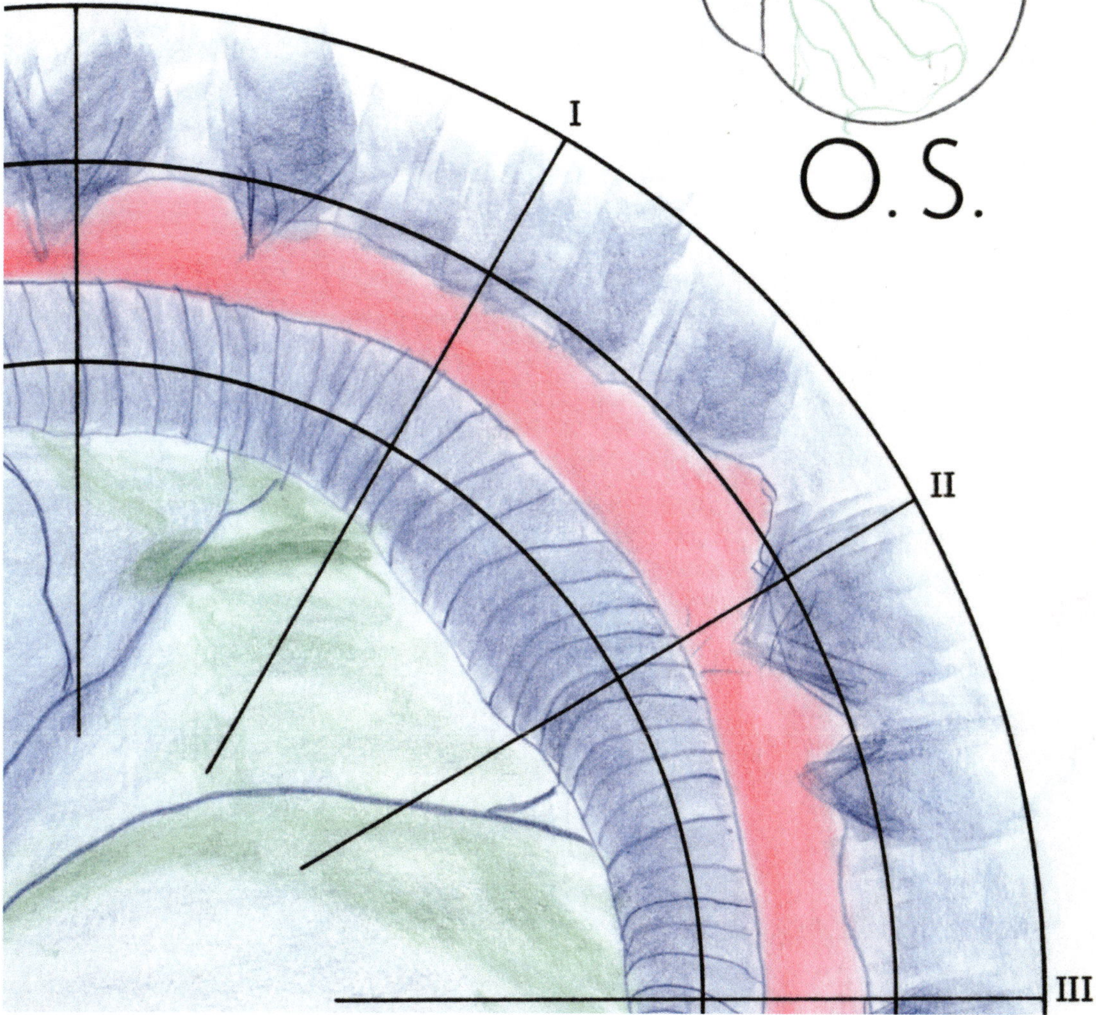

Shown here is a giant retinal tear of about 180 degrees, at the point where it may flop over onto itself. At the time of this drawing, the tear would have been difficult to unroll. The choroid, which is the layer under the retina, is drawn in delicate radial orange lines that are oriented to the direction of the vessels. Since the center of vision would not be affected yet, this patient likely had good vision but was on the precipice of sudden, and likely permanent, loss.

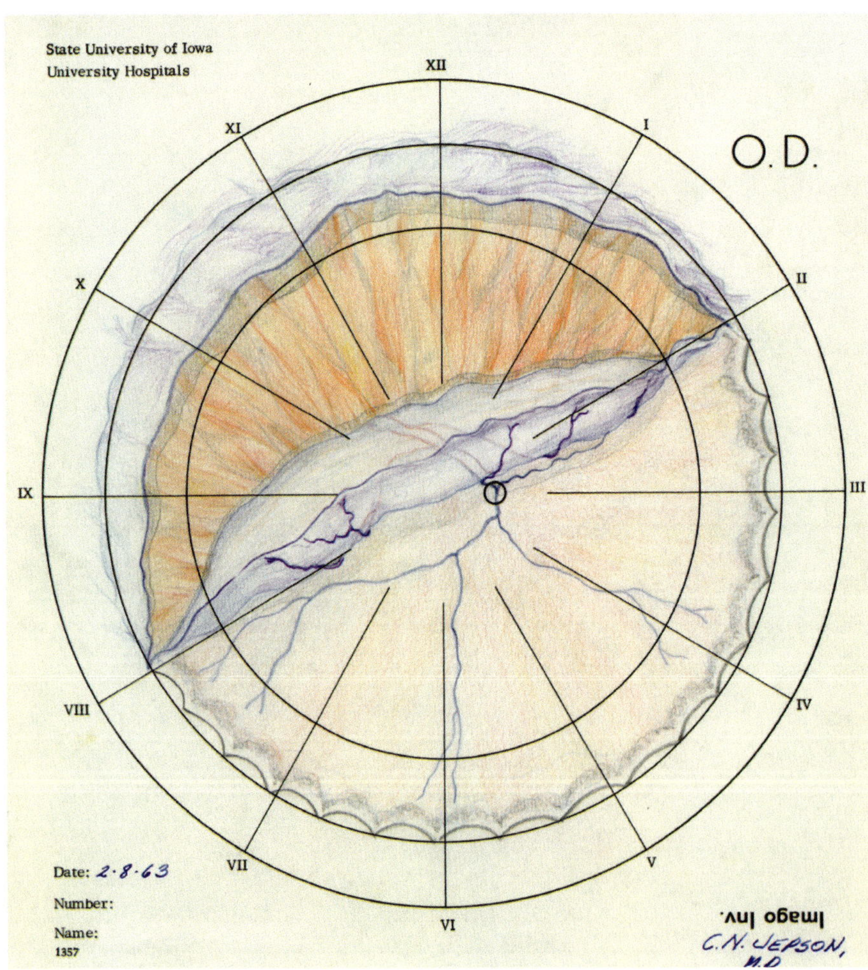

Artist C. Neal Jepson
February 8, 1963

Diagnosis:
Traumatic retinal detachment, right eye: "struck in the right eye while working on a car as a mechanic. Now has a giant retinal tear of whole superior fundus 180 degrees."

Given the inferior location of this giant tear, the prognosis was very poor. One technique available was to use a Stryker roll frame to rotate the patient's head down in hopes of flattening out the rolled edge of the inferior flap.

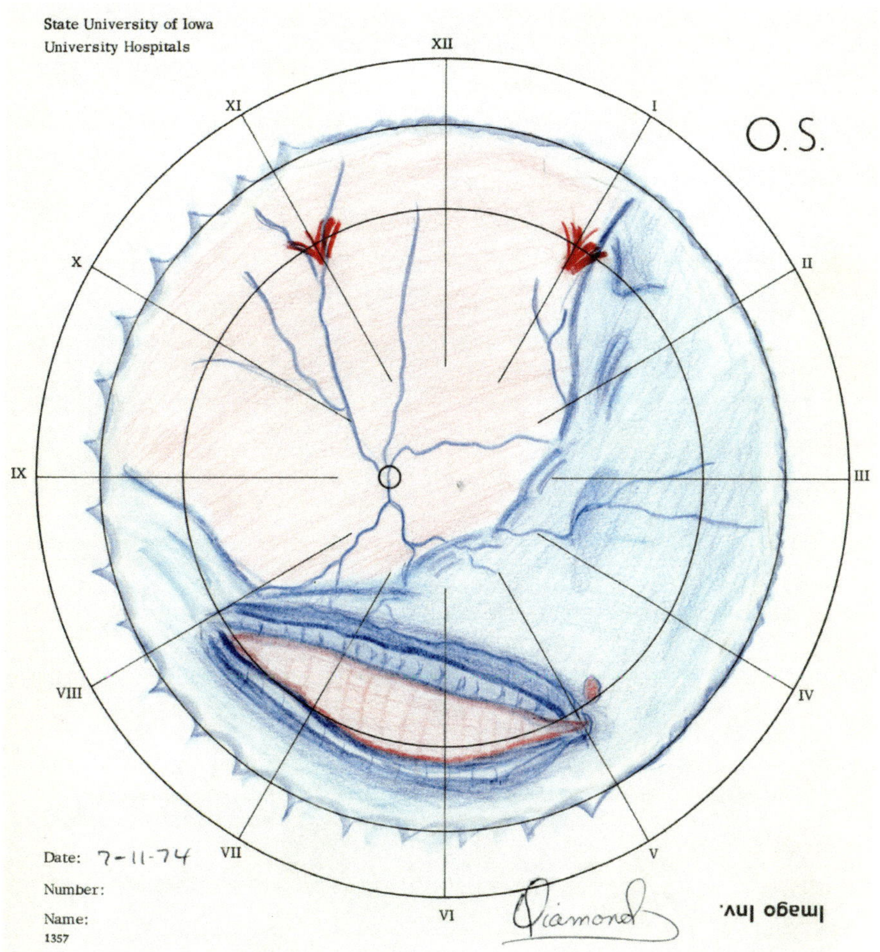

Artist James G. Diamond
July 11, 1974

Diagnosis:
Inferior giant tear after a "penetrating injury to the left eye resulting from a retained thorn"; cataract extraction one month earlier.

When they are drawing details temporally, or on the temple side of the eye, examiners have difficulty seeing over the patient's nose, especially if they are short. The artist likely worked quickly in rendering this giant tear.

"Since I am only 5'2" I used to stand on a stepstool to draw."

Charlotte Burns, MD (6/24/11)

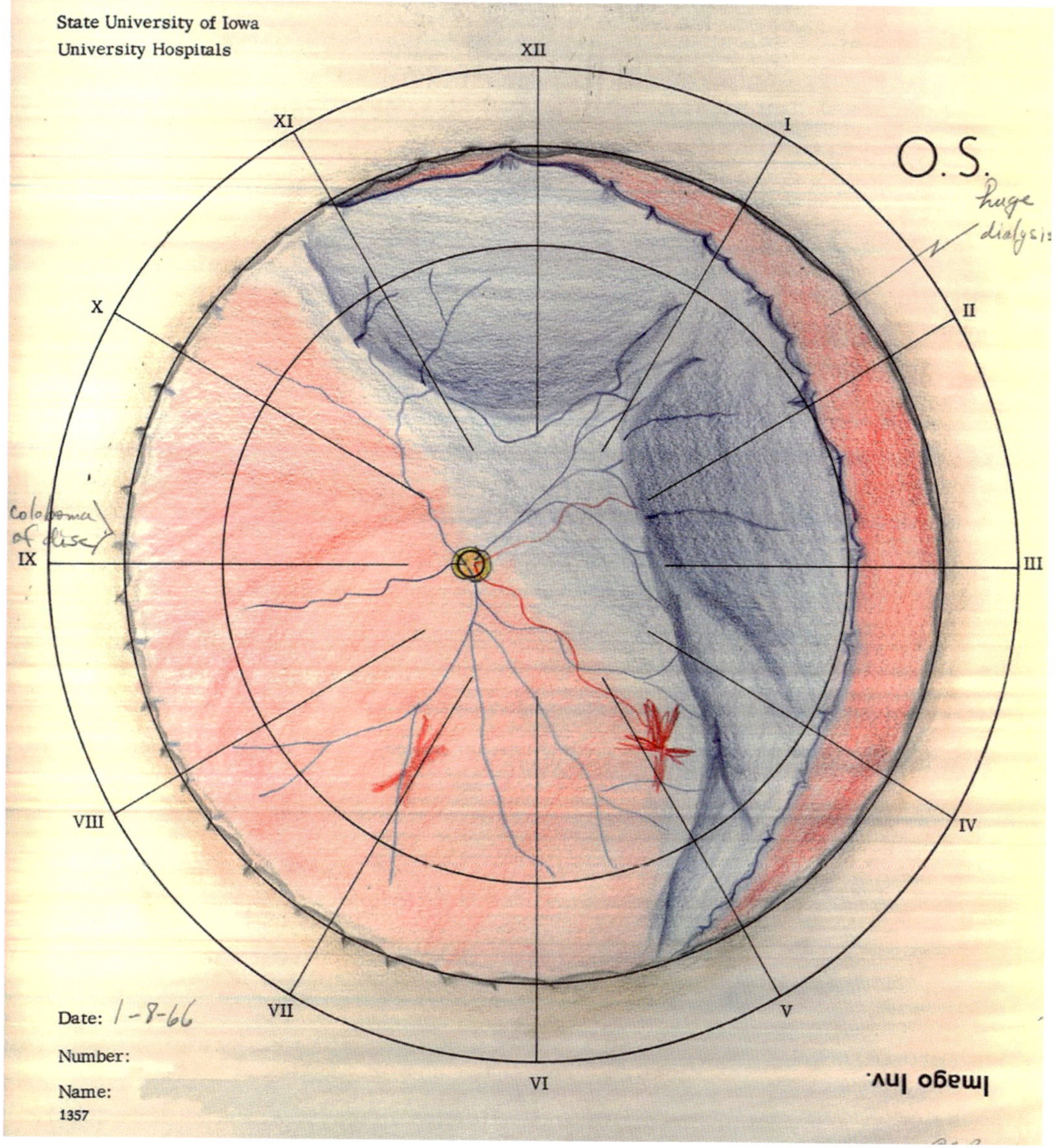

Artist Charlotte A. Burns
January 8, 1966

Diagnosis:
Large disinsertion (dialysis) of the retina, left eye; degenerative myopia.

"I thought long and hard about [this patient]," Dr. Ed Ferguson wrote in 1960, "and finally came to the conclusion that in my hands this is an inoperable detachment. It is difficult for me to give up on a 20/50 eye." Because the patient was still functionally sighted in that eye and because vitrectomy was not yet an option, Dr. Ferguson suggested that the patient see Dr. Paul Cibis in St Louis, who had "a method of introducing a needle within the eye and evacuating fluid underneath the retina by means of this needle and actually sucking the retina against the opening of the needle and attempting to incarcerate the retina in the hole of the sclera."

"Unfortunately," Dr. Cibis wrote in a letter to Dr. Ferguson, "the bridge which your drawing shows between twelve o'clock and one o'clock had come loose when the patient arrived in St. Louis." Then, after providing a full clinical explanation of how he "had hoped to avoid further progression of the detachment," Dr. Cibis added, "The retina has come loose more, and I feel that any further attempt of surgery is due to fail."

This is a beautiful and elegant drawing of multiple retinal tears, one of which is a giant tear. Almost resembling an orchid, this image is a dramatic representation of traumatic retinal injury. Although giant retinal tears have been observed in the setting of no antecedent trauma, the large horseshoe-shaped tear in the left lower corner is diagnostic of globe trauma.

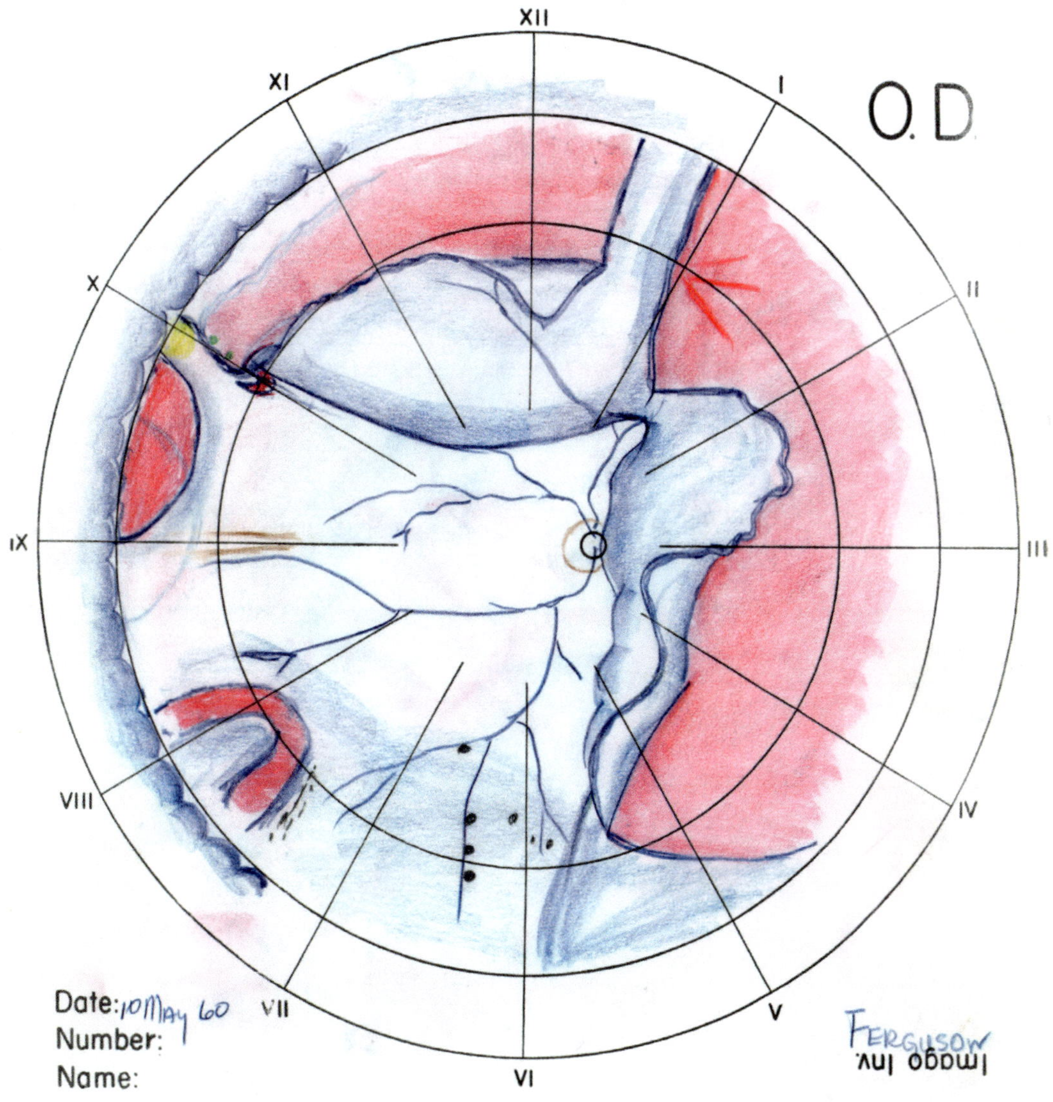

Artist Edward C. Ferguson
May 10, 1960

Diagnosis:
Almost complete peripheral detachment and two large areas of dialysis.

According to Dr. Burton (7/7/11), the drawing shows a parachuting retina, such that "if you'd place a rose bowl into a fruit bowl (the choroid), we'd see what that drawing shows. The retina is still attached at the back...and collapsing."

Understanding the full implication of this terribly serious, complete (360-degree), giant retinal tear clashes with the beauty of the artist's rendition. The ring of red with bright, branching choroidal vessels uncovered by the retina highlights the torn edges of the retina, indicated by the fine, dark-blue, zigzag lines. The back portion of the retina, which is shown to be attached (red center), has the edges of the retinal tear folded over it like the petals on a closing flower.

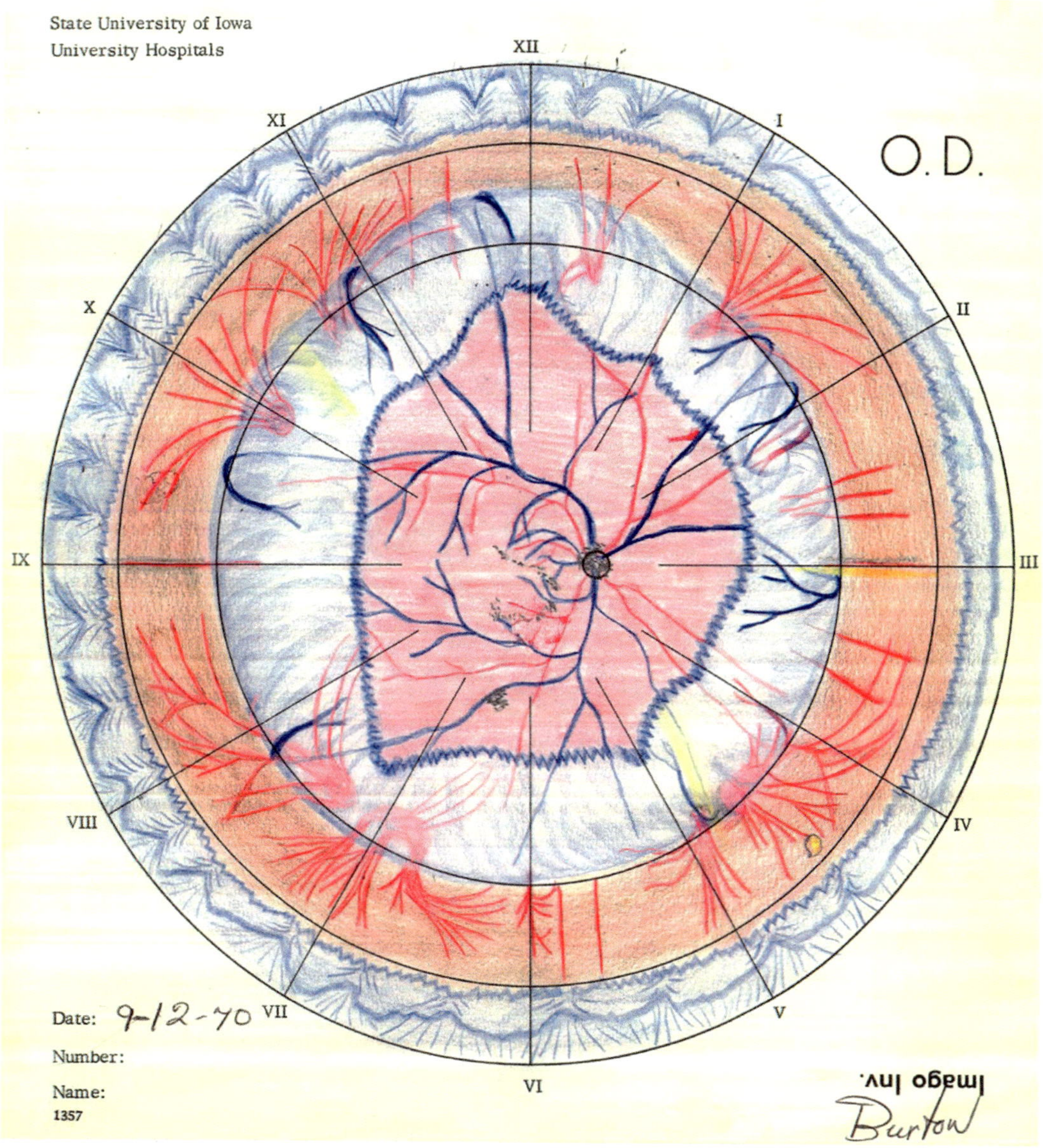

Artist Thomas C. Burton
September 12, 1970

Diagnosis:
Traumatic retinal detachment with 360-degree giant tear, right eye (had been struck in that eye with a hairbrush two weeks earlier by a two-year-old).

Cystoid Degeneration

The most forward or peripheral portion of the retina is thinner and more cellularly simplified than the central retina, which is responsible for central vision. A common feature of the peripheral retina is the development of fine cysts between layers of the neural retina. These may be present in regions or around the circumference of the retina, as shown here.

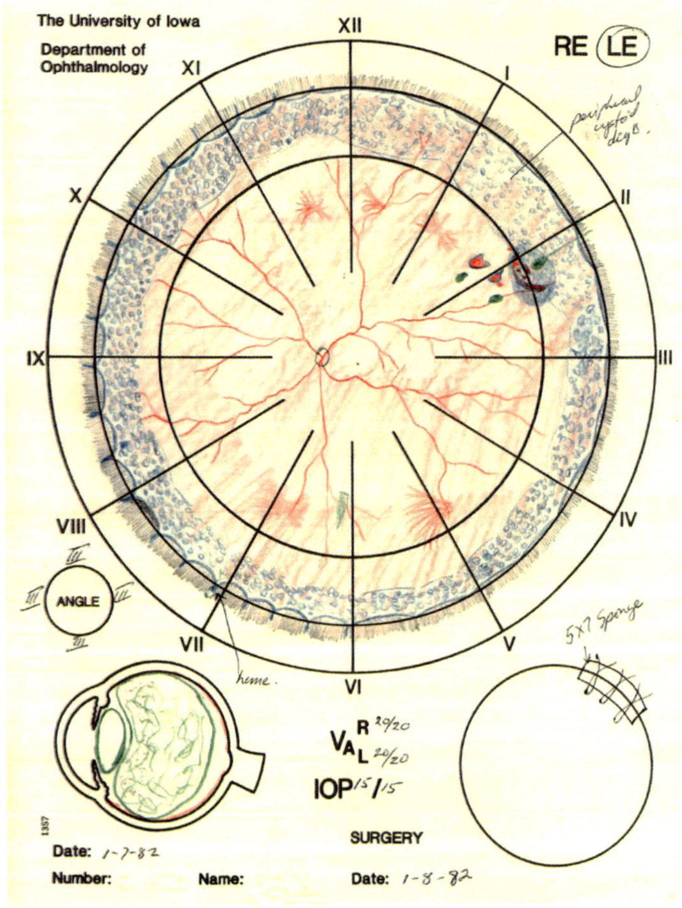

Artist Michael J. Howcroft
January 7, 1982

Diagnosis:
Superotemporal flap tears and operculated hole, left eye.

When a retinal detachment is long-standing, the detached retina may undergo cystic degeneration with formation of a very large retinal cyst, confirming the tendency of the peripheral retina to form cystic structures. This unusual drawing shows that the patient had both peripheral cystoid degeneration and a large retinal cyst.

"[There is an] interplay between real, impressive, and admirable people who explored, expanded, and brought this part of retinal therapy into a new level of success… all sharing individual drive to excel not only in the pathology seen when examining, its accurate representation, but all having the inner muse that brought these drawings to an exceptional and varied art form."

Bob Baller, MD (8/17/11)

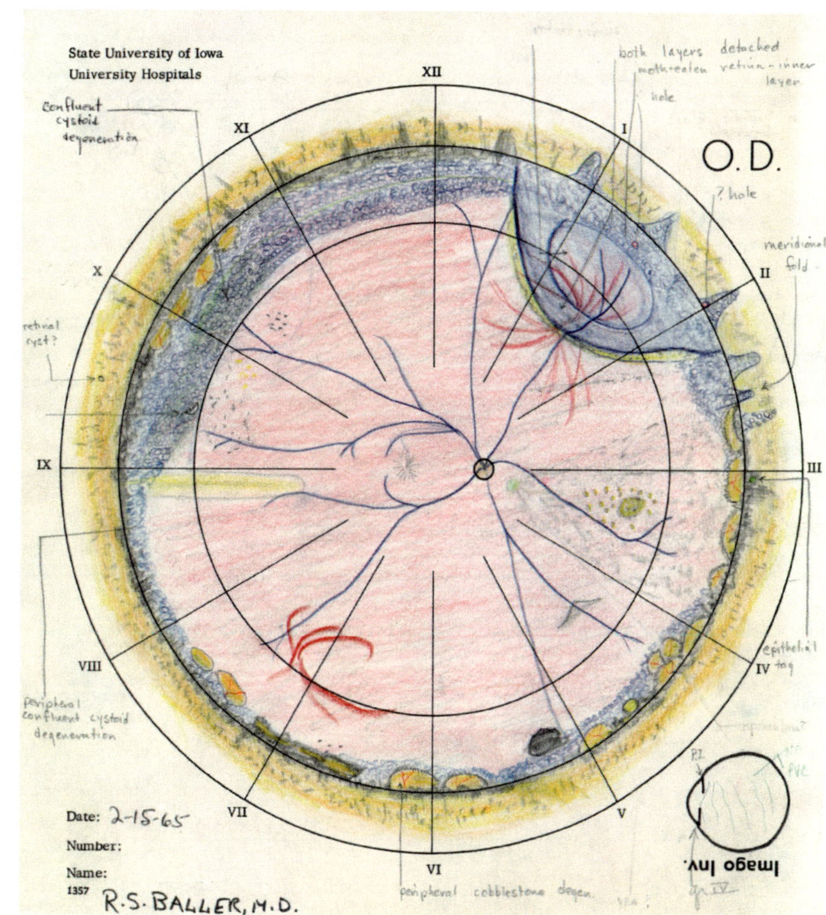

Artist Robert S. Baller
February 15, 1965

Diagnosis:
Retinal detachment, right eye, and cataract extraction three years earlier.

Peripheral retinal cysts in an attached retina typically occur in confluent regions. In rare instances the cysts may protrude inwardly. This is defined as a cystic retinal tuft, and it may mimic retinal tears. In this drawing, isolated cystic changes, likely representing these tufts, are shown. (Peripheral cystoid degeneration has smaller and confluently distributed cysts.)

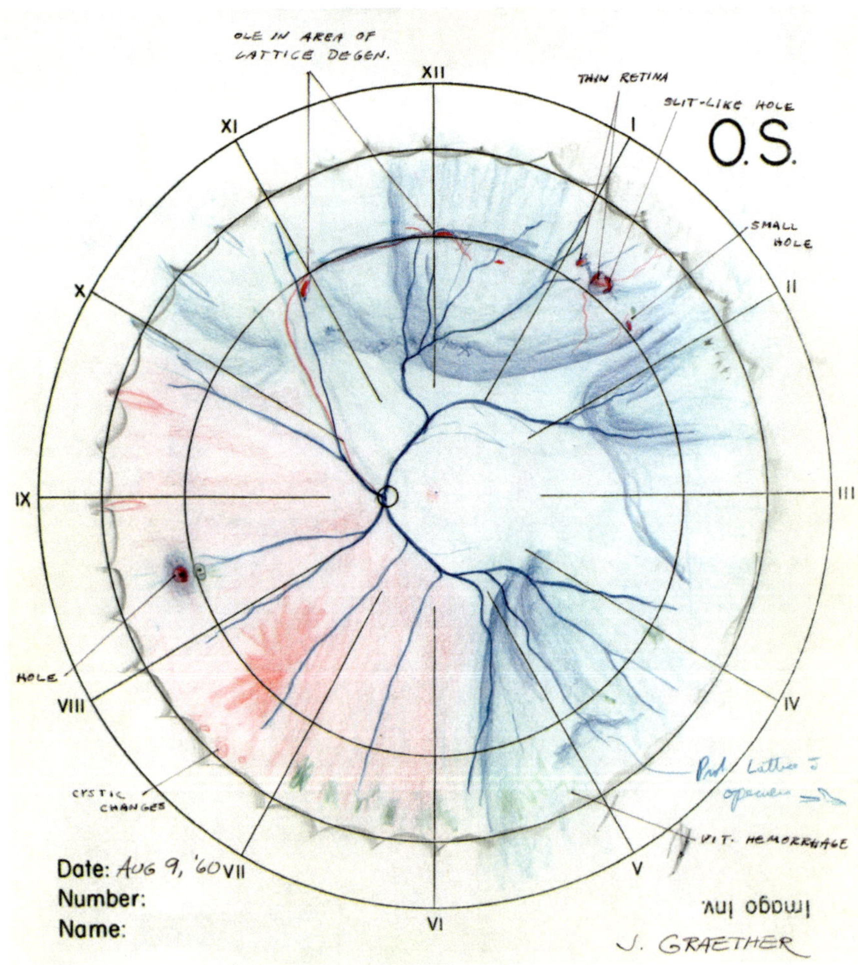

Artist John M. Graether
August 9, 1960

Diagnosis:
Retinal holes, lattice degeneration, and a macula that is flatly off. Visual symptoms over the prior six weeks had been "red flashes like fireflies...then spots appeared...and then a cobweb appeared."

The peripheral cystoid degeneration present in this patient is shown as a uniformly wide band around the peripheral retina and is unrelated to the location of the retinal tear. Peripheral cystoid degeneration is not believed to convey a risk of retinal detachment.

"Fundus drawing was a wonderful way for the residents to study retinal. It was labor intensive, well worth the effort. Dr. Watzke was instrumental in how important the drawings worked as a learning device; in the operating room you were able to locate anatomic details from the drawing—a great help."

Charles E. Jones, MD (1/13/11)

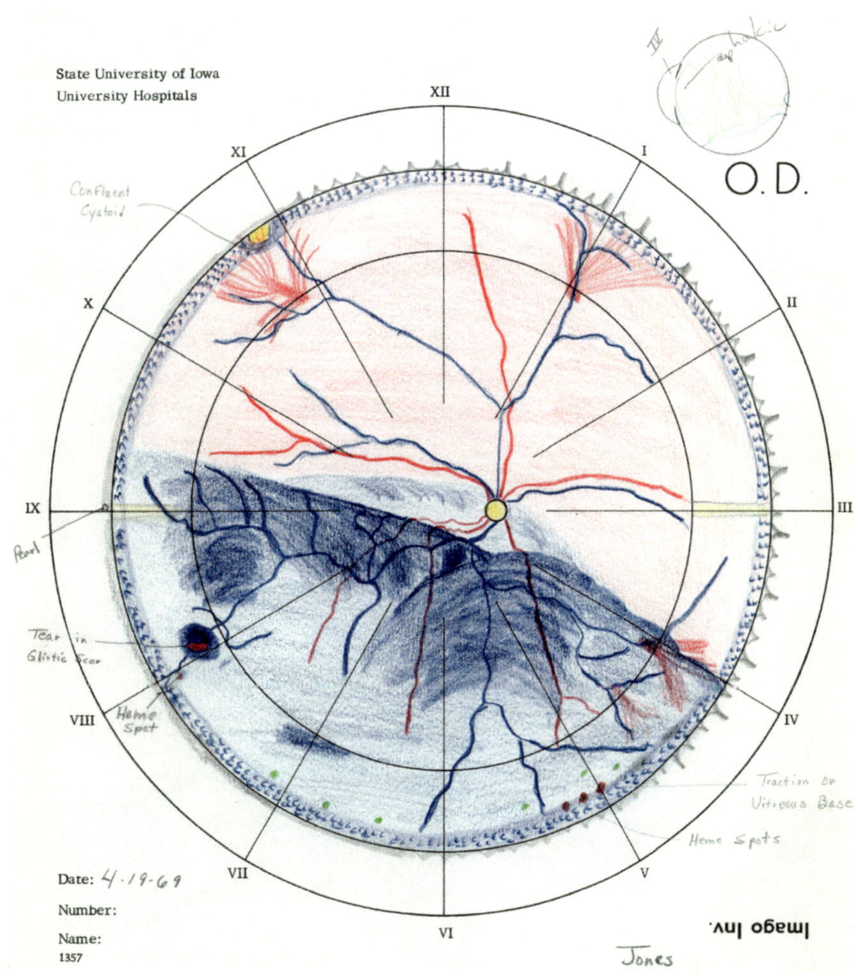

Artist Charles E. Jones
April 19, 1969

Diagnosis:
Retinal detachment, right eye, following a cataract extraction one year earlier.

A similarly distributed but pigmented peripheral band at the edge of the retina is illustrated here. Pigment is usually found beneath the retina in the retinal pigmented epithelium or deeper choroidal layer.

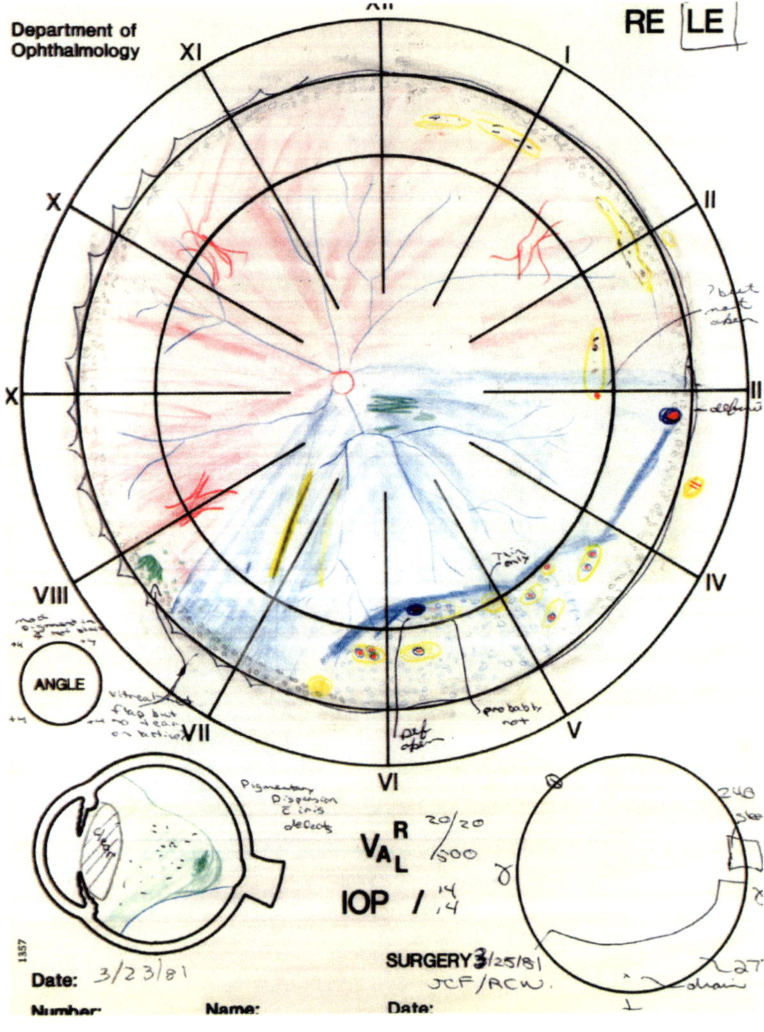

"Messy is a style"

Steve Russell, with respect to Jim Folk's drawings

Artist James C. Folk
March 23, 1981

Diagnosis:
Retinal detachment, left eye, and pigmentary dispersion type syndrome.

This patient had both peripheral cystoid degeneration and peripheral pigmentation. The interesting texture rendered within the cystoid region is a particularly nice shading effort that illustrates the undulating folds that may be present over the area of retinal detachment.

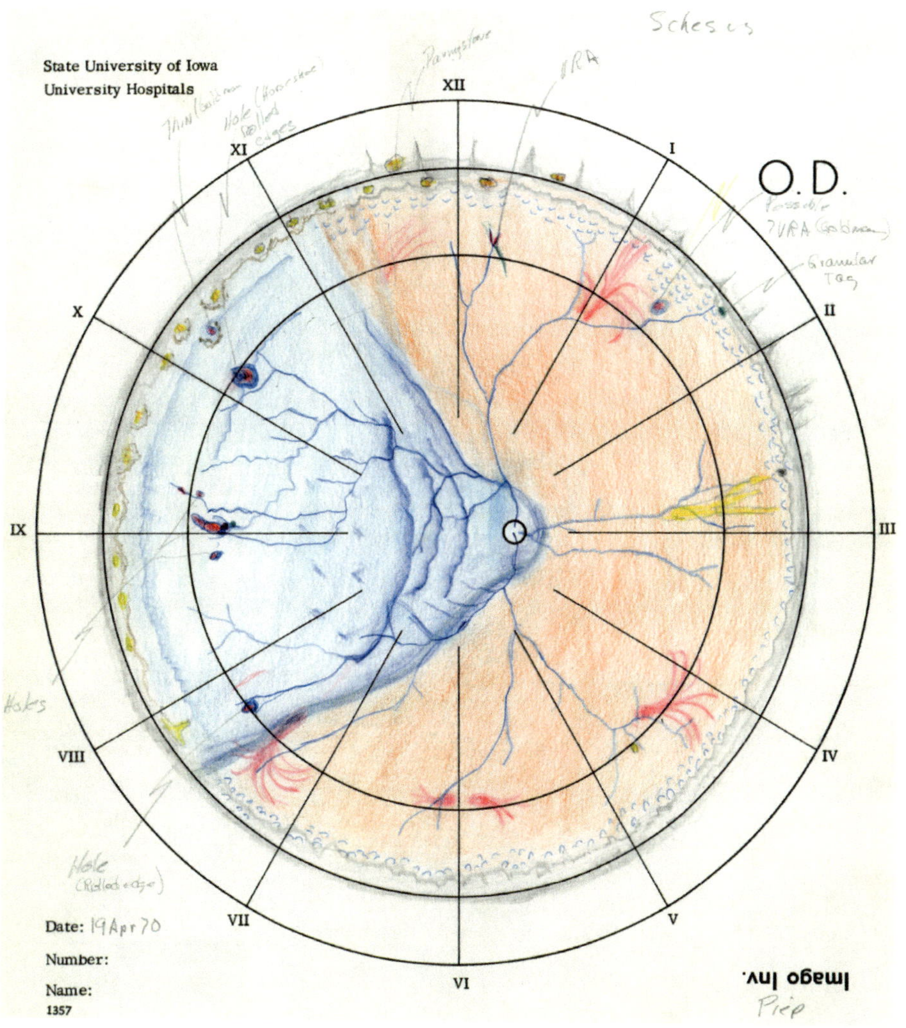

Artist Larry G. Piepergerdes
April 19, 1970

Diagnosis:
Retinal detachment, right eye.

The dramatic nature of this complicated drawing is clear. This patient had a retinal detachment over a prior scleral buckle. The anterior retina and ora serrata have been detailed with extraordinary care. Areas of peripheral retinal and choroidal atrophy are shown throughout the bottom half of the drawing, likely representing previous cryotherapy, or freezing treatment. Based on the date of the drawing, however, the atrophy might also represent regions of prior diathermy treatment.

"Looking at Dr. Jepson's drawing, it almost seems like the view through a modern wide-angle viewing system; he was a true artist!"

Patrick Caskey, MD (4/20/11)

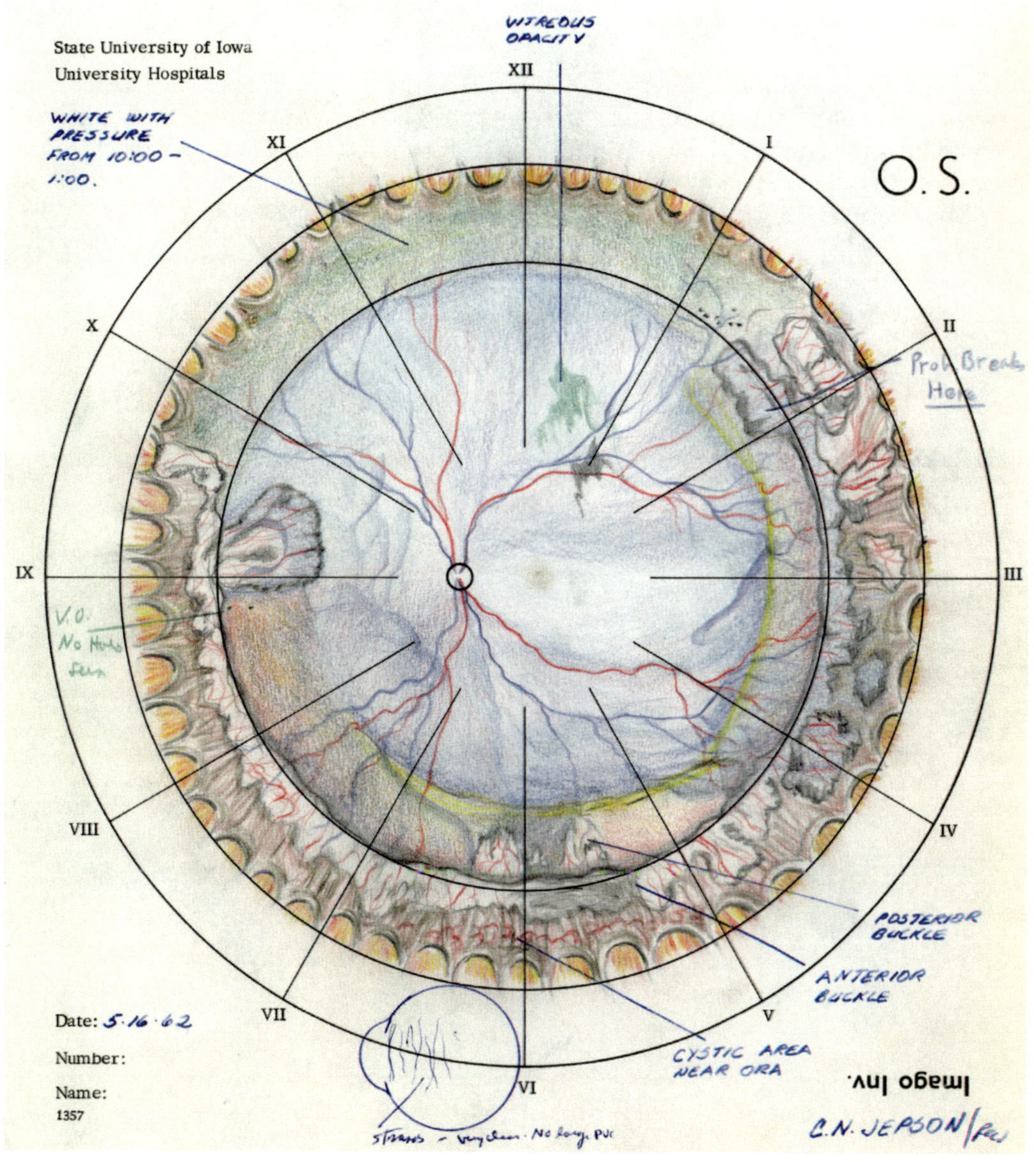

C. Neal Jepson
May 16, 1962

Diagnosis:
Retinal detachment, left eye, with previous cataract extraction.

Bullous Detachments

With a completely different stylistic flair, this drawing illustrates small intraocular gas bubbles seen clustered together in green with a reflection on each bubble. Intraocular gas is used to position and hold the retina and to block fluid from entering through retinal holes, which the bubbles cover. Note that the retina billows like a sail or is elevated like a blister, a characteristic referred to as a bullous retinal detachment.

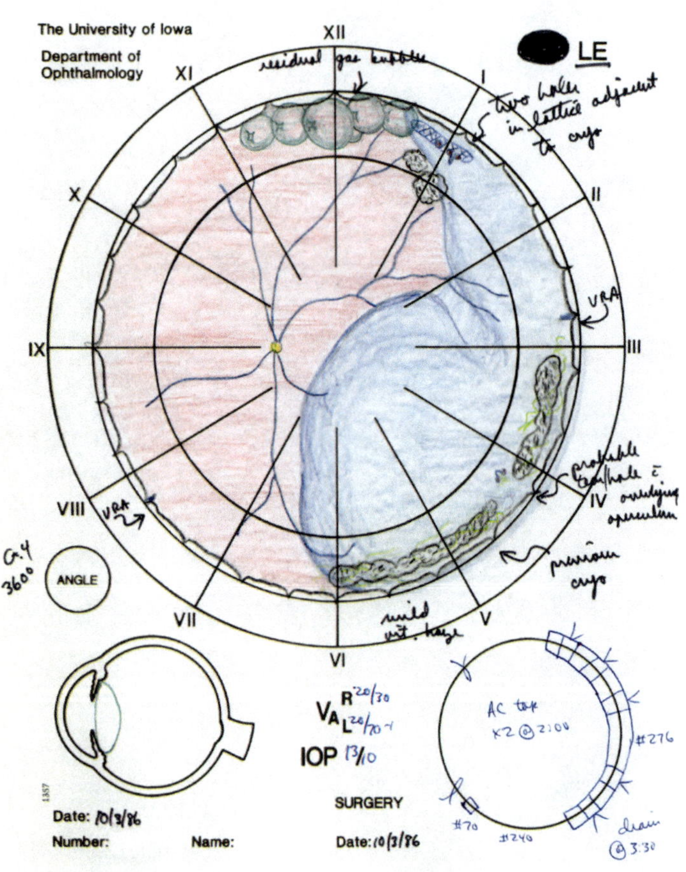

"The color drawings are representations of the retinal pathology I saw as I was evaluating the patient for surgery. The smaller notations below and to the right are drawings and descriptions of what was done at the time of surgery to correct the retinal problems."

Pat Caskey, MD (12/6/10)

Artist Patrick J. Caskey
October 3, 1986

Diagnosis:
New area of peripheral fluid and two new peripheral breaks adjacent to an area of lattice degeneration in the left eye, following pneumatic retinopexy three weeks earlier for a retinal detachment and following cataract extraction one year before the detachment.

Here is another beautifully complicated retinal drawing that contains peripheral retinal cystoid degeneration and a similar bullous retinal detachment, and shows the attached retina over a scleral buckle. The retinal detachment extends from the holes between one and two o'clock via a tunnel above and left of the retinal detachment.

"[Drawing in the way colleagues draw] does add perspective but [does] not necessarily influence the treatment program."

David Fenske, MD

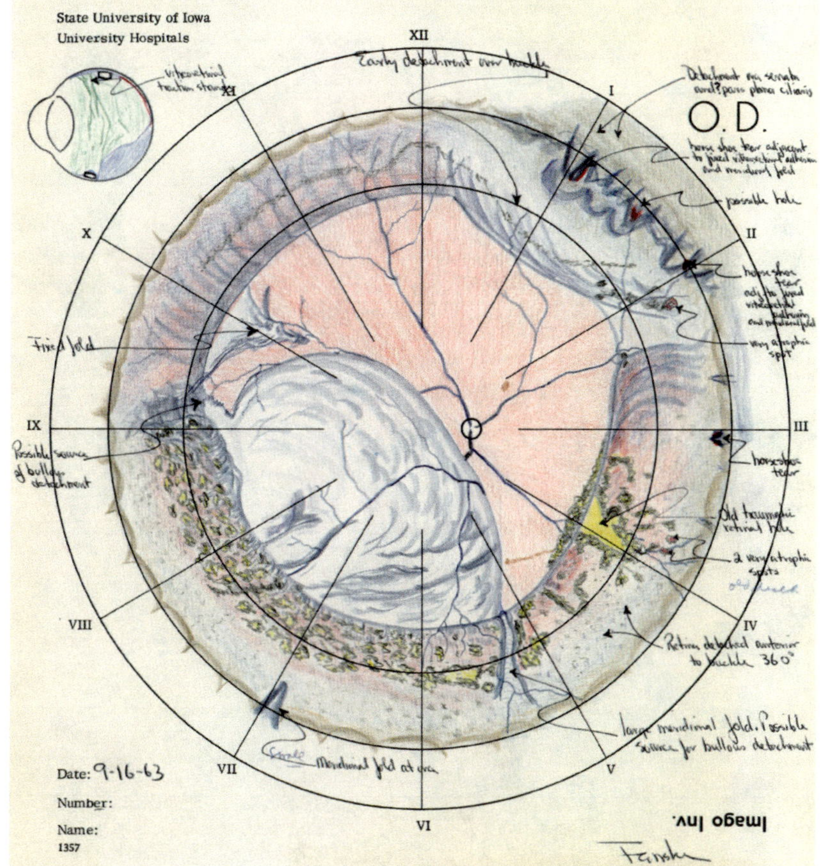

Artist David Fenske
September 16, 1963

Diagnosis:
Recurrent retinal detachment, right eye, following trauma one year earlier (had been hammering metal with metal, and a piece flew into the right eye); had two previous repairs.

Here the unremarkable curvature of the inferior retina is shown in contrast to the brightly hued superior billowing retinal detachment. The shading was achieved with lines and an eraser, which is unusual. Most shading is done with color addition. The artist included peripheral cystoid degeneration and depicted lattice degeneration in the lower right in an unusual way.

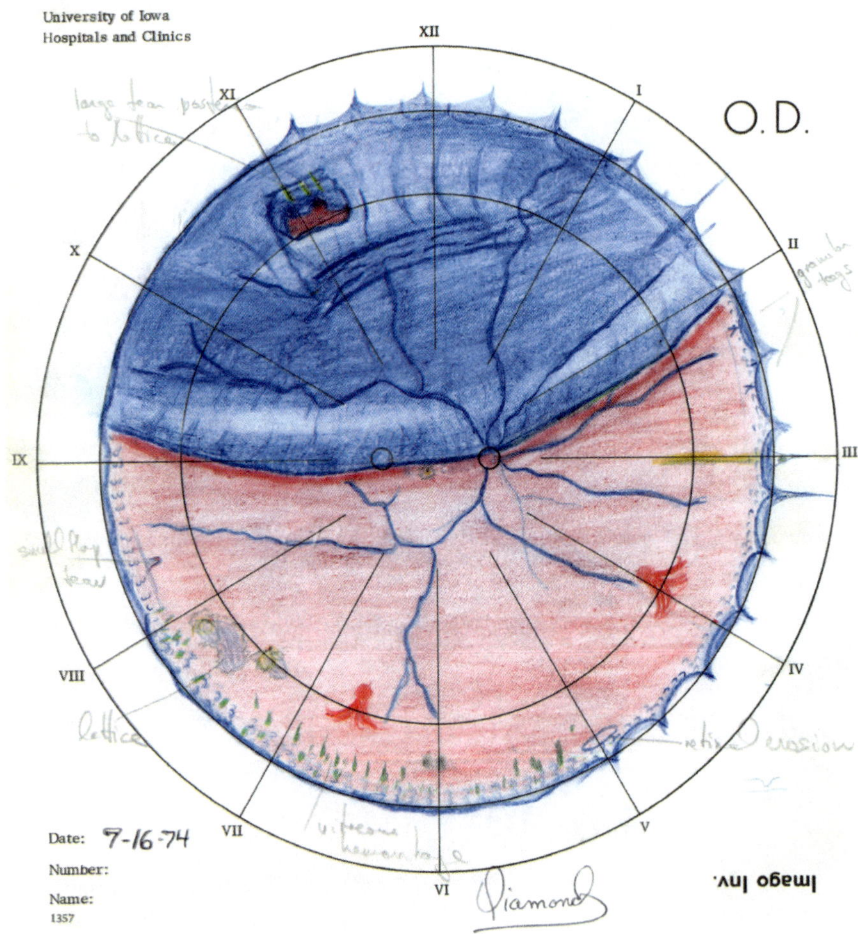

Artist James G. Diamond
July 16, 1974

Diagnosis:
Large bullous retinal detachment, right eye, with an extremely large horseshoe tear.

In this strikingly detailed rendering of the retinal arteries (red lines) and veins (blue lines), the artist has increased the drama by adding choroidal pigment parallel to the retinal vessels (shown in light black). The pigment between the choroidal vessels is virtually never included in drawings, as it is not useful as a landmark because they appear so similar to each other.

"Those of us who spent an hour or two (or longer) drawing the retina of the detachment patient most certainly had a more intimate sense of the patient and his (her) problem than do the present-day residents. If they don't spend time looking into the eye and drawing the retina, how do they represent the case and its possible therapy? I'd like to know how they do it in this day and age."

Neal Jepson, MD (6/12/11)

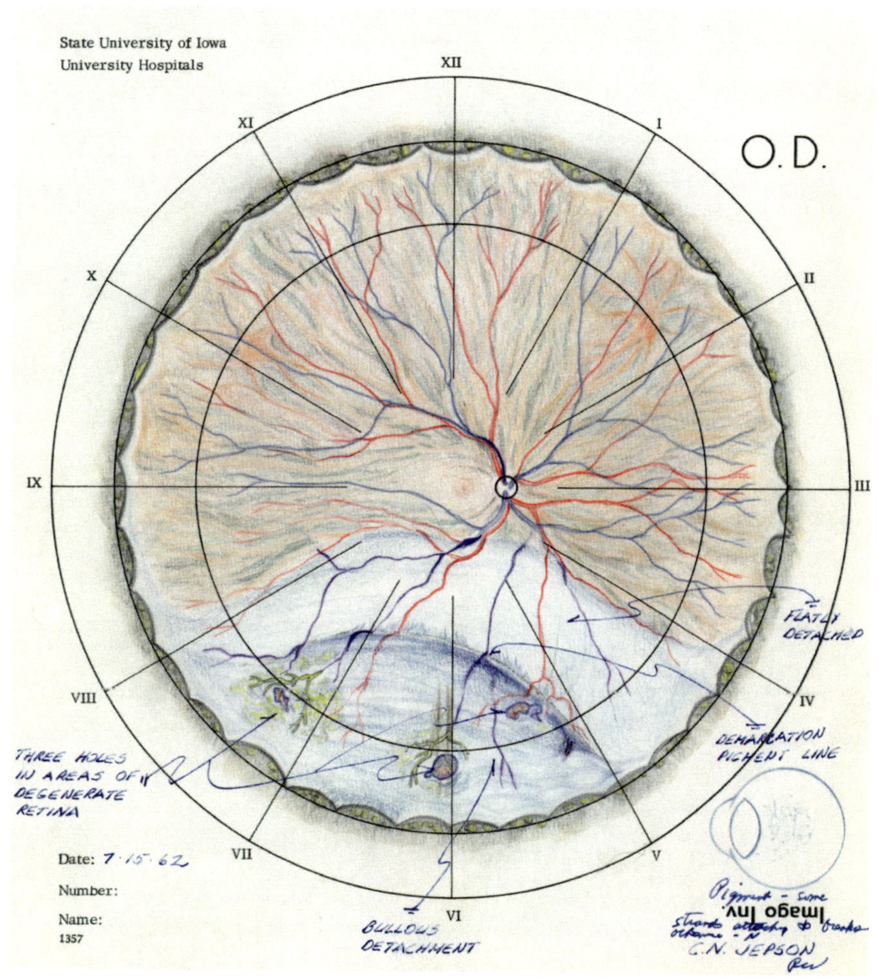

Artist C. Neal Jepson
July 15, 1962

Diagnosis:
Bullous retinal detachment, right eye (had experienced a stiff neck before right upper field loss of vision).

As a preview to a later chapter on the ora serrata, this drawing is shown with a retinal detachment extended to include a bullous detachment including the pars plana (shown in blue at the top). Another anomaly is that the billowing retina overhangs the optic disc. This feature will sometimes cause the vision to be reduced without detachment of the macula or central retina.

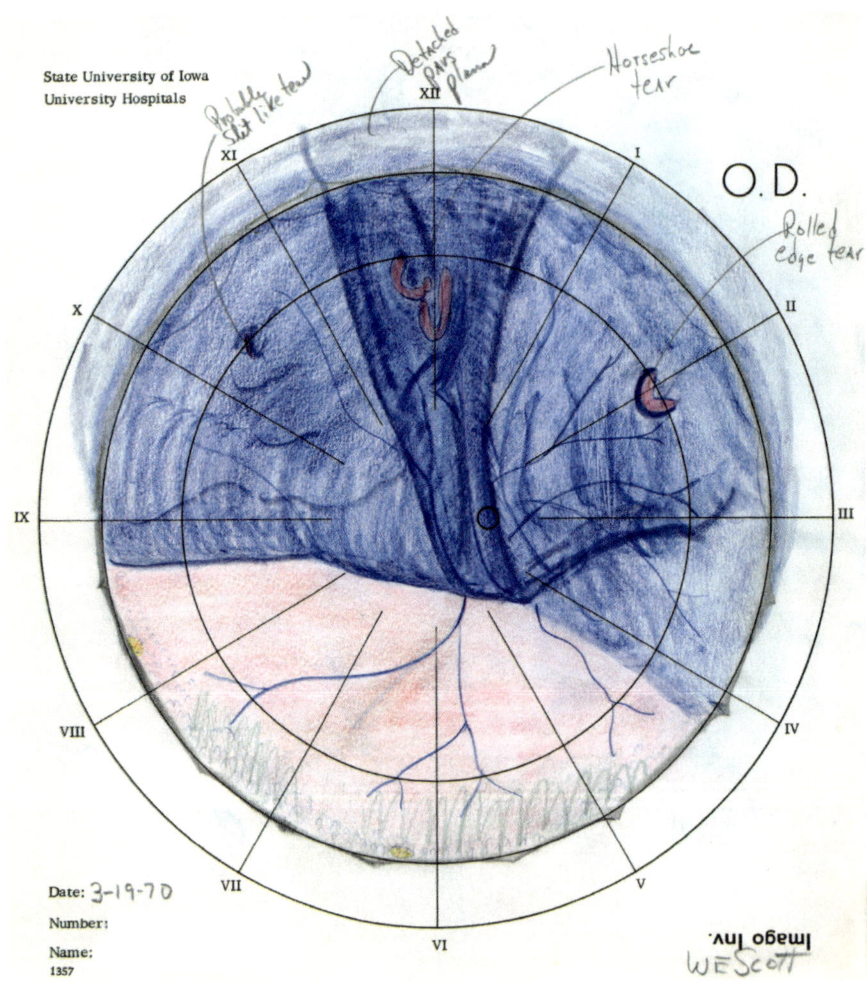

Artist William E. Scott
March 19, 1970

Diagnosis:
Large superior tear with bullous detachment, right eye.

In a common precursor to a giant retinal tear, this patient's row of retinal tears is at risk of extending, like perforation holes in paper. Unusual and occasionally ambiguous coloration is used for the attached and detached retina.

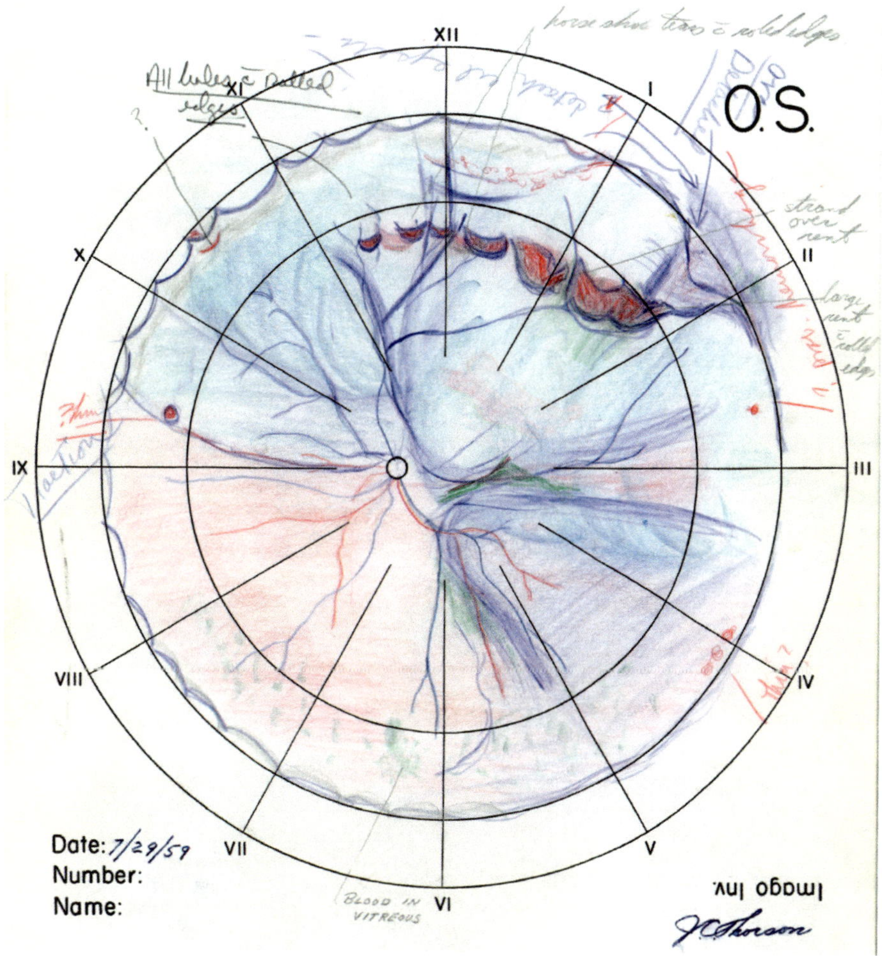

Artist Jon C. Thorson
July 29, 1959

Diagnosis:
Bullous retinal detachment, left eye.

Rarely seen in the collection, the shadow that the elevated retinal detachment casts on the adjacent attached retina is illustrated by the bright-red shading along the bottom and right of this detachment. The aqua blue used for the detachment contrasts with typical blue of the color code, giving this a pastel appearance. Here the detachment billows over the optic disc, hiding it, just as in the previous drawing.

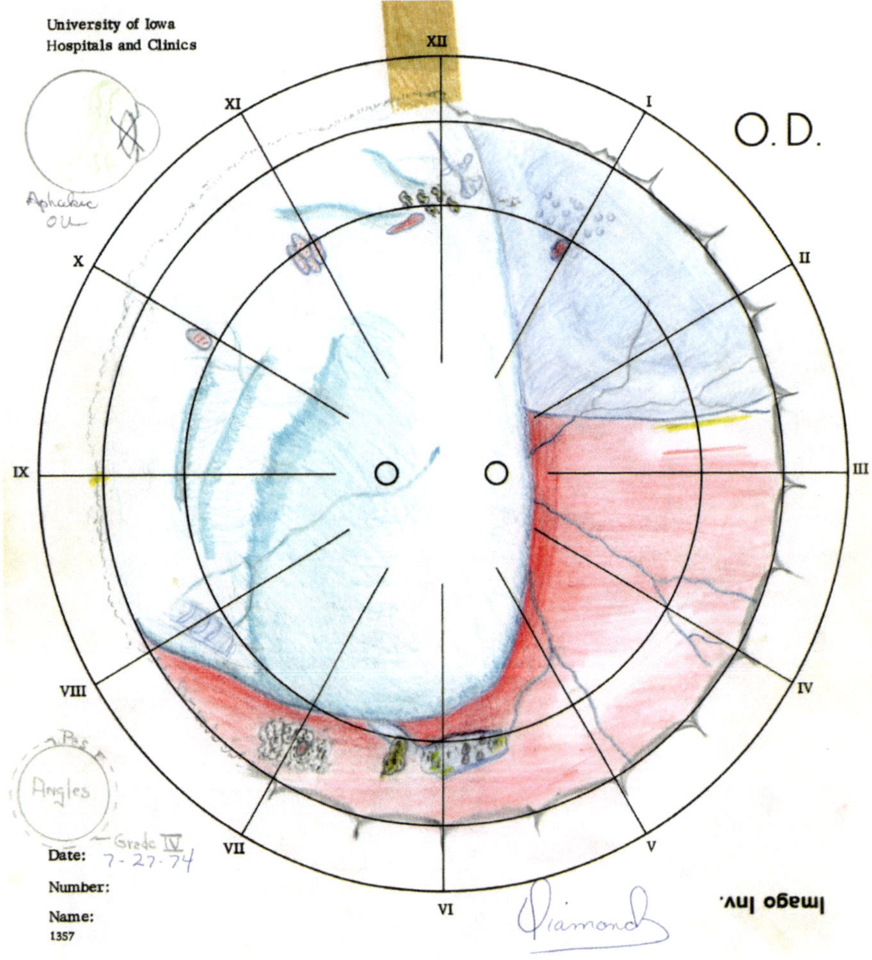

Artist James G. Diamond
July 27, 1974

Diagnosis:
Bullous retinal detachment, right eye, with two large tears superiorly; cataract extraction two years earlier.

Cysts

A cyst is a fluid-filled sac and is, when associated with the retina, a rare occurrence. In this wonderfully delicate depiction of a large retinal cyst, the combination of red and blue gives an illuminating and impressionistic feel to the drawing. Although the retinal cyst suggests the detachment had been present for a long time, the small red spots near the top suggest this patient had retinal dialysis, a type of detachment likely to be repaired.

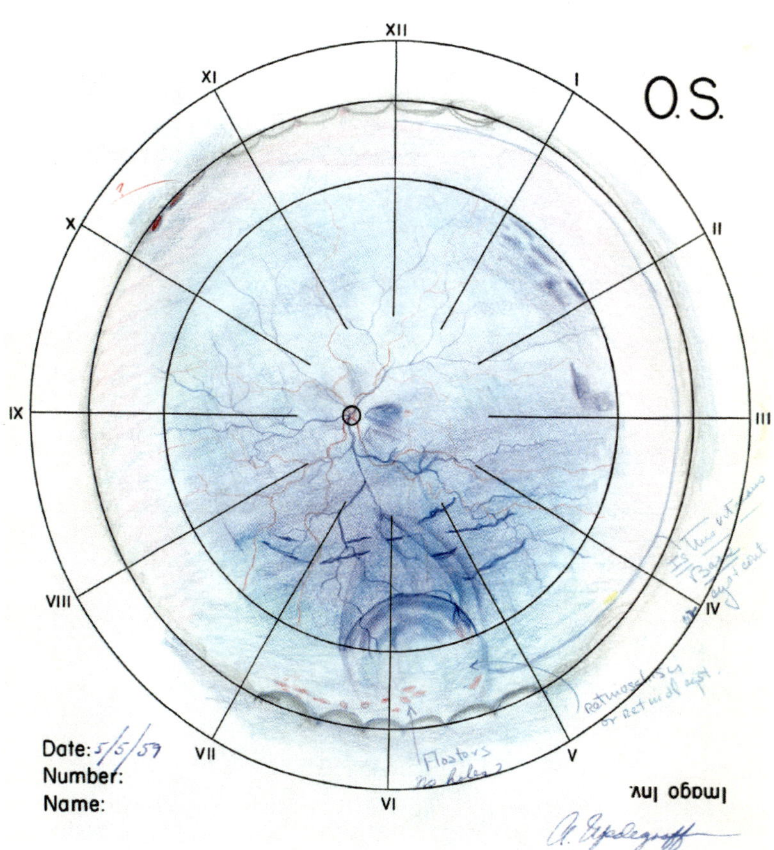

Artist Ambrose G. Updegraff
May 5, 1959

Diagnosis:
Traumatic retinal detachment from retinal dialysis, left eye, with large retinal cyst following a football injury six years earlier with subsequent loss of vision (injury had been to the left malar process with retinal edema about the macula and persistent retinal edema and folds temporal to the macula).

In this depiction of a large retinal cyst, the center is less blue, as many cysts have thin walls and may be nearly transparent. Interestingly, this patient also has a macular cyst.

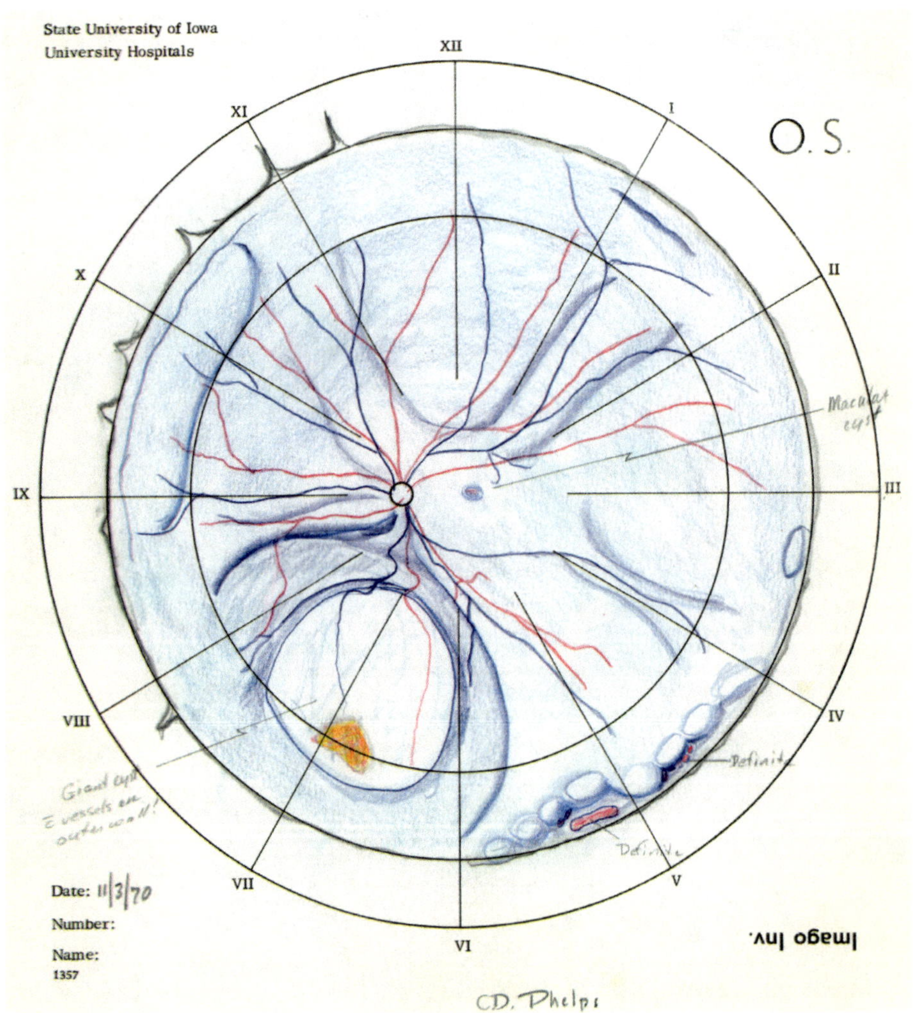

Artist Charles D. Phelps
November 3, 1970

Diagnosis:
Total retinal detachment, left eye, with an infra-temporal dialysis and several large infra-retinal cysts; had previous trauma (hit by a rock two years earlier with progressive loss of vision).

Multiple retinal cysts due to retinal detachment suggest that the entire retina may be atrophic, limiting the possibility that the retina could be reattached. Adding a sense of spontaneity, the artist completed the shading in blue and blue-purple with fluid undulations.

"Drs. Watzke and Ferguson believe this eye to be essentially inoperable."

(Illegible signature, 12/8/60)

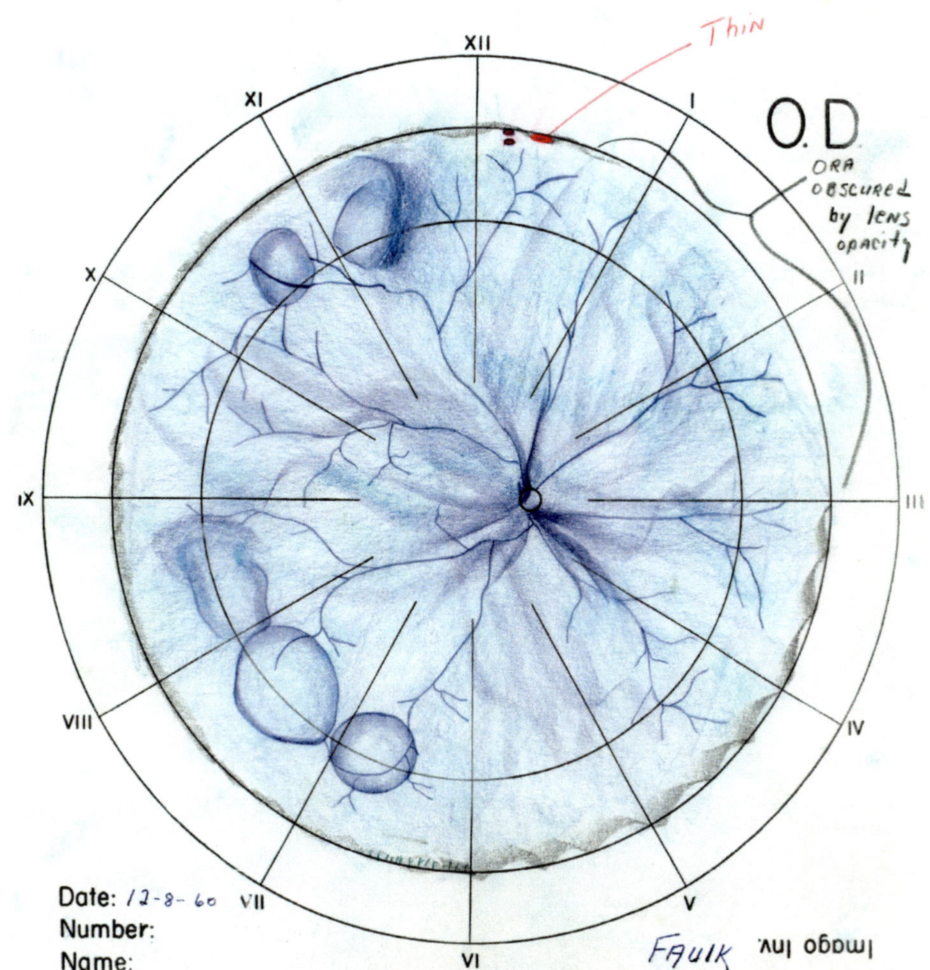

Artist Wallace H. Faulk
December 8, 1960

Diagnosis:
Complete bullous retinal detachment, right eye; fell from a truck in 1954 and had subsequent retinal detachment in 1955; also had a cataract extraction earlier.

Star Folds

The most prominent feature in this drawing is the herringbone shading, used to illustrate fixed interior folds on the retina. Where these folds come together, they form a star-shaped pleat called a star fold, which is an indication of the degree of inner retinal cellular proliferation termed proliferative vitreoretinopathy, or in the era of this drawing "massive periretinal proliferation."

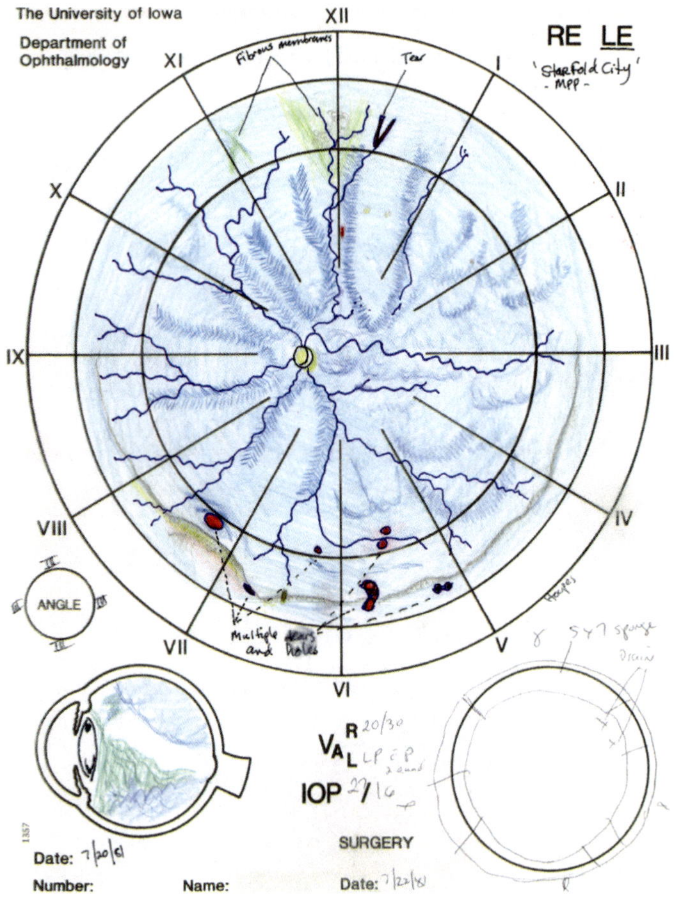

"Star Fold City"

Phillip Hoopes, MD (7/20/81)

Artist Phillip C. Hoopes
July 20, 1981

Diagnosis:
Retinal detachment with massive periretinal proliferation, left eye.

When pleats or star folds form in a detached retina, the folds may appear U-shaped as shown here. This situation was further complicated by the presence of asteroid hyalosis, in which numerous small, white calcific nodules are distributed within the vitreous, preventing a crisp view of the retina.

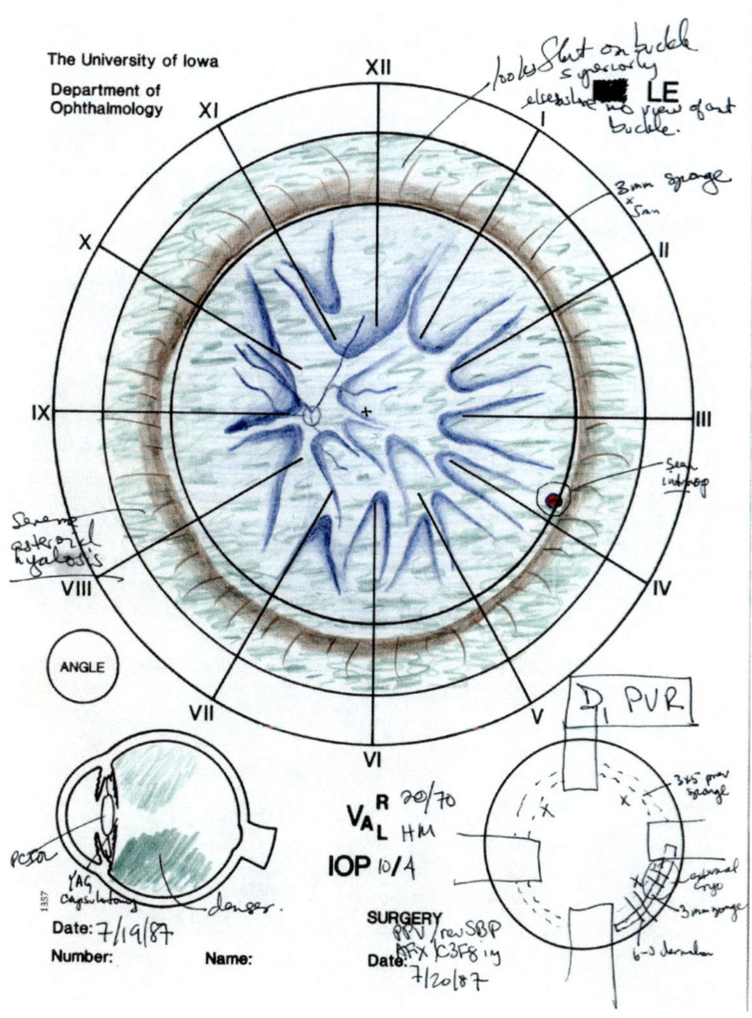

Artist Stephen R. Russell
July 19, 1987

Diagnosis:
Recurrent rhegmatogenous retinal detachment, left eye, with moderate to severe proliferative vitreoretinopathy, early star folds in all quadrants, and dense asteroid hyalosis; cataract extraction nine months earlier.

The degree of traction or force on the retina is sometimes inferred from the contour of the retinal vessels. On the left side are several vessels that have more tortuosity, or coiling, than normal vessels (a normal one is shown near twelve o'clock). On the left side are illustrated the fine retinal folds characteristic of surface cellular proliferation and membranes termed proliferative vitreoretinopathy.

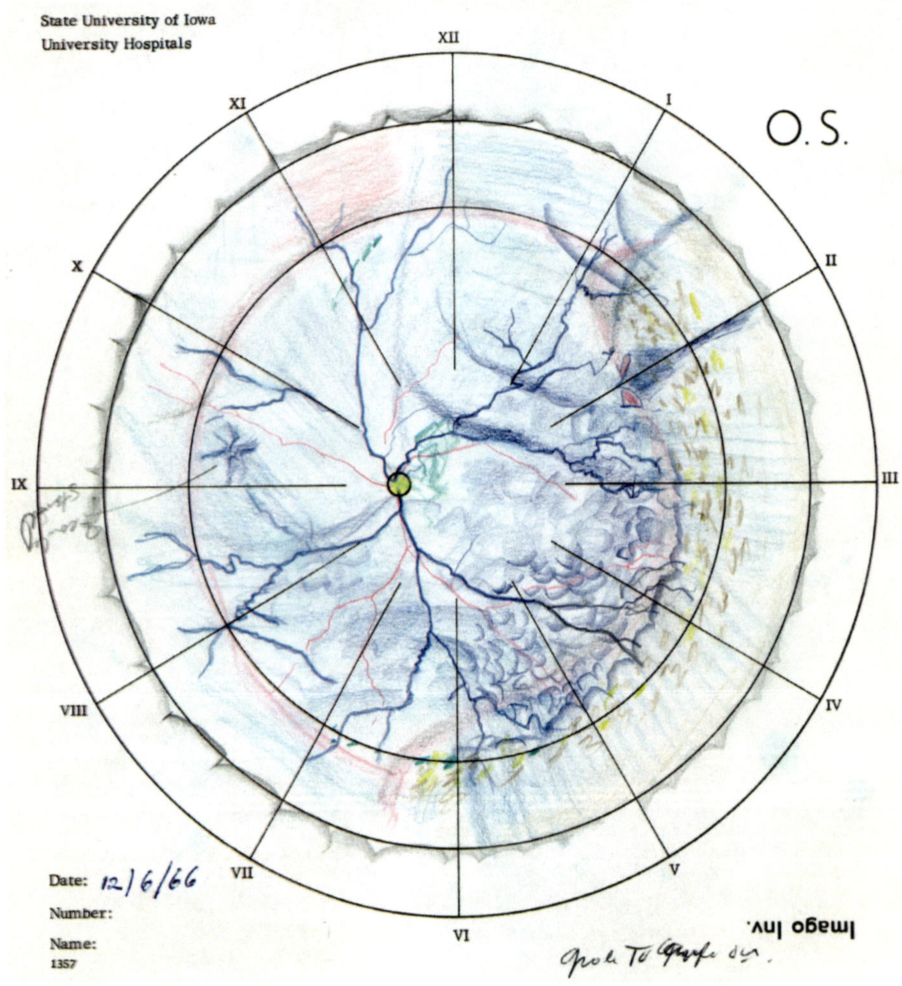

Artist Donald J. Doughman
December 6, 1966

Diagnosis:
Recurrent retinal detachment, left eye.

At the top is neatly shown an attached retina within the bend of the scleral buckle. At the bottom, a typical location for proliferative vitreoretinopathy, is shown the tightly folded retina near the optic disc and the contraction folds along the inner retinal circle, the anatomic location of the vitreous base.

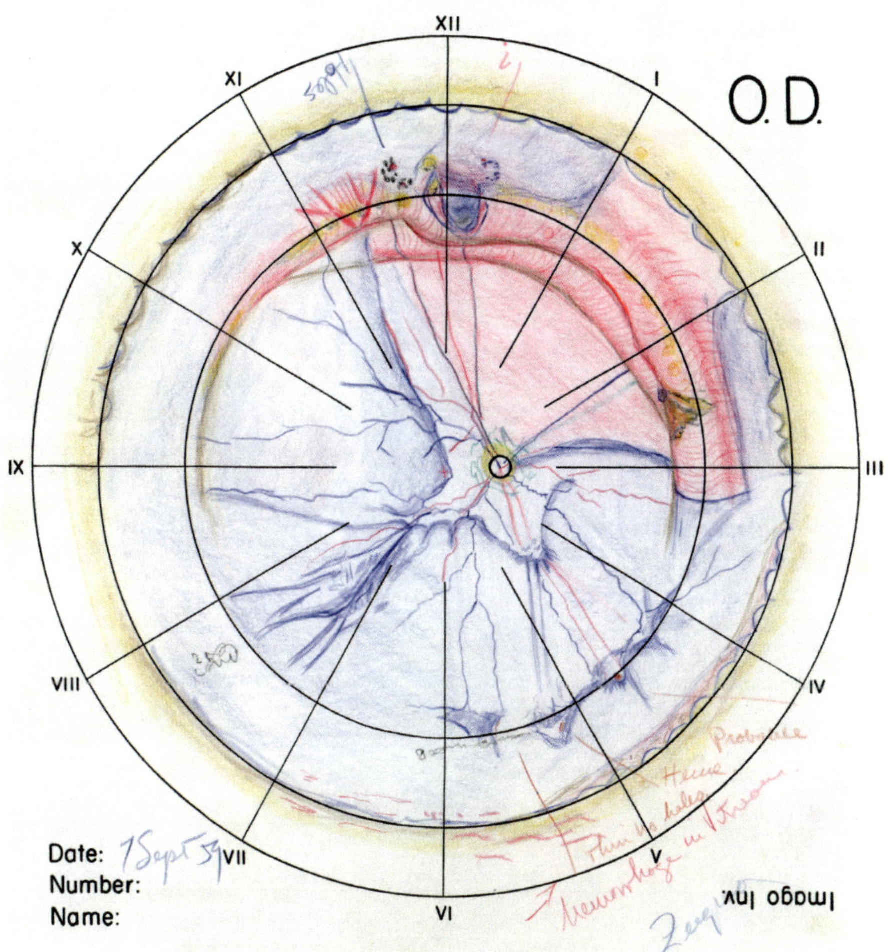

Artist Edward C. Ferguson
September 7, 1959

Diagnosis:
Recurrent retinal detachment, right eye.

Choroidal Detachments

When serous fluid or blood accumulates within or beneath the choroid, this material elevates the choroid from the scleral coat of the eye, resulting in a choroidal "detachment." To the left, top, and right in this drawing are choroidal detachments, which act much like a scleral buckle to keep intraocular fluid from passing through a retinal hole (one shown here at twelve o'clock). Also complicating this patient's problem is the retinal detachment shown at the bottom, which could be a preexisting, hole-related retinal detachment (rhegmatogenous retinal detachment) or a more recent one due to the choroidal detachment (exudative retinal detachment).

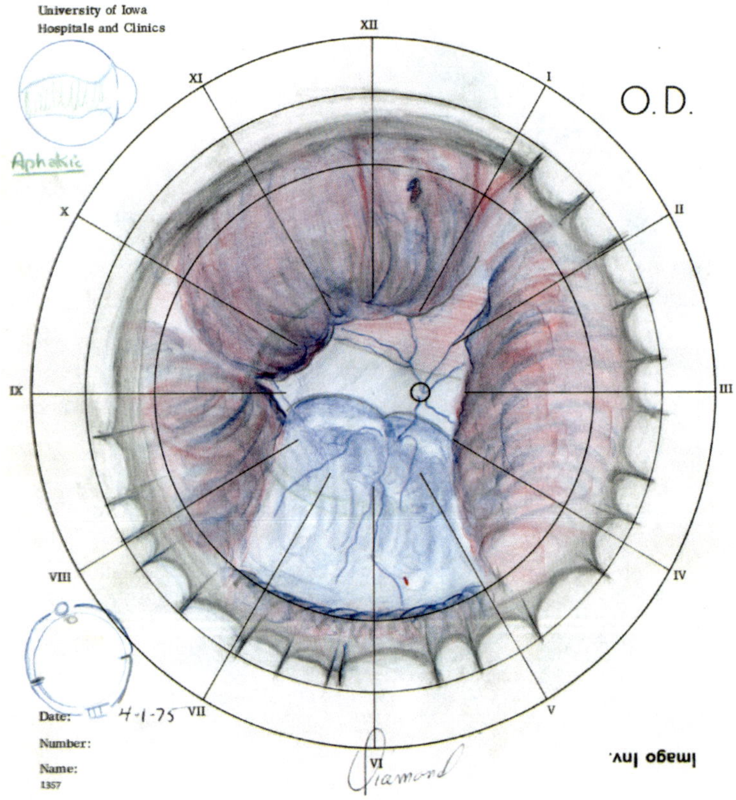

Artist James G. Diamond
April 1, 1975

Diagnosis:
Longstanding retinal detachment with choroidal detachment, right eye; cataract extraction six weeks earlier.

In this seemingly simple drawing, the choroidal detachment is shown on the bottom left, and it extends back as a blue triangular area beneath the retinal detachment. Two radial retinal folds are drawn at ten and twelve o'clock.

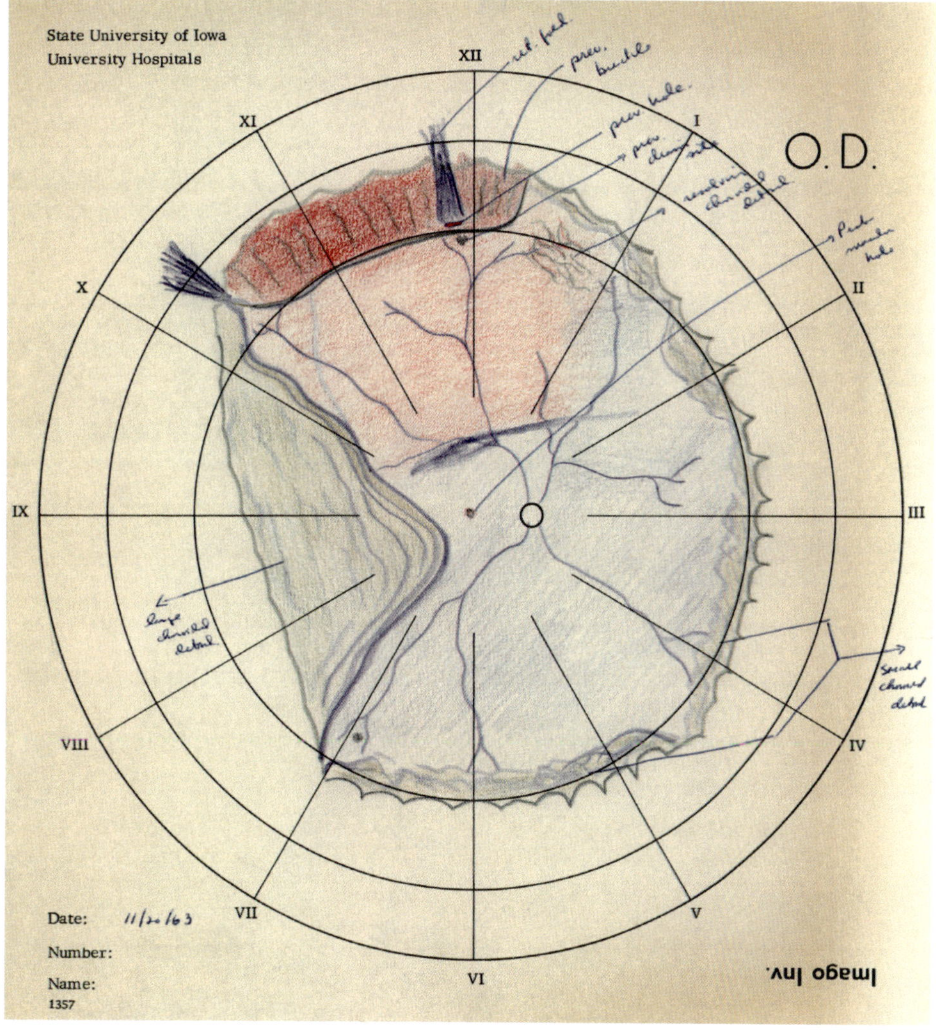

Artist James M. Hersey
November 20, 1963

Diagnosis:
Recurrent choroidal detachment, recurrent retinal detachment, right eye, and previous cataract extraction, right eye.

Concentric colored rings belie the complexity of this drawing. The blue retinal detachment, seen in the center, is surrounded by billows of green representing blood in the vitreous. The outer scalloped rings of brown are the mounds created by choroidal detachments that are caused by bleeding into the choroid.

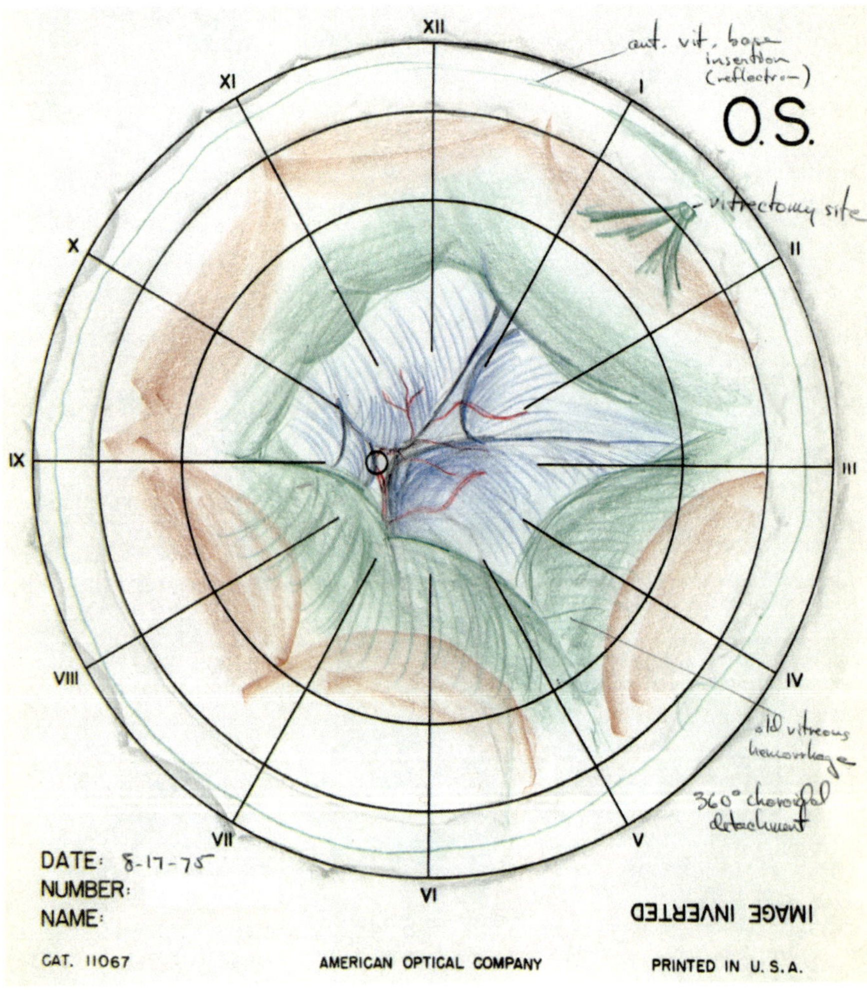

"I created lots of drawings that, in my opinion, were of much better quality. But perhaps you need this to demonstrate a special condition."

Tom Burton, MD (12/10/10)

Artist Thomas C. Burton
August 17, 1975

Diagnosis:
360-degree choroidal detachment, left eye, following cataract extraction and subsequent vitreous hemorrhage.

Asymmetric choroidal detachments are exaggerated by the perspective distortion of different regions of the ora serrata and the radiating brown shading over the face of the detachments. Shown in the depths of the eye is the retina, detached below. In addition, the retinal veins appear to have increased tortuosity or wiggles due to the perspective when viewing the retina at a shallow angle.

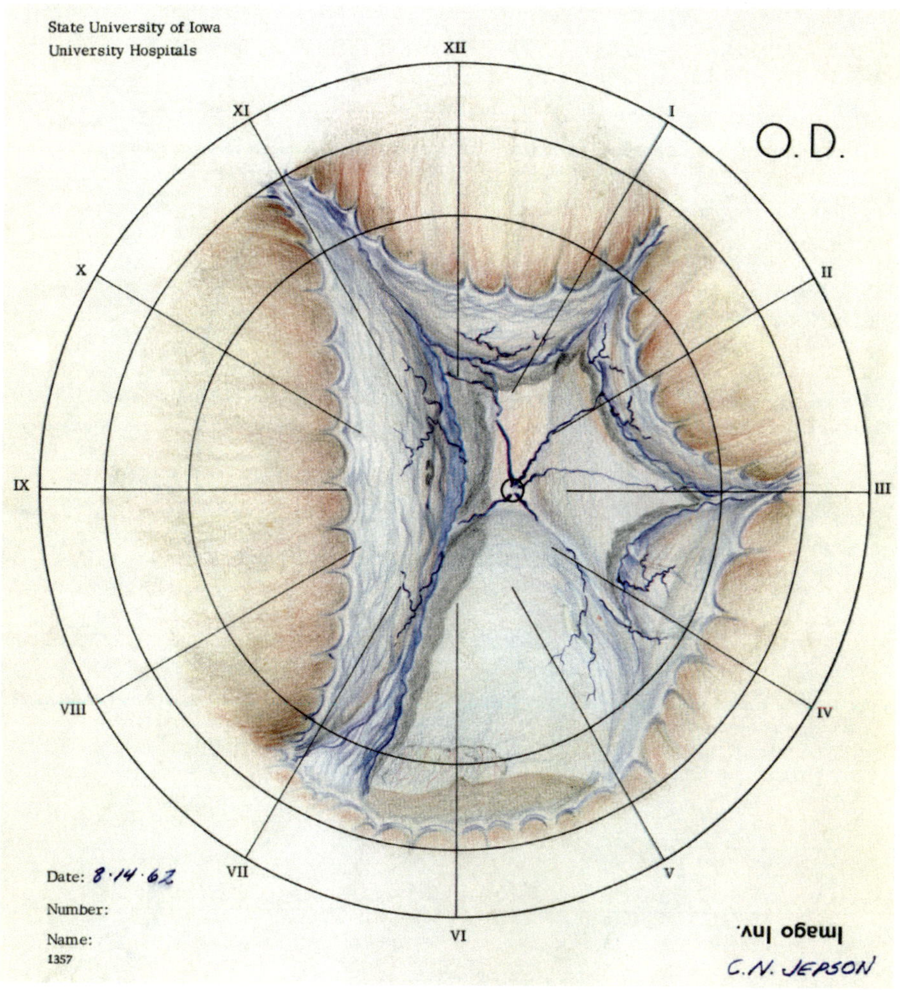

Artist C. Neal Jepson
August 14, 1962

Diagnosis:
Retinal detachment, right eye, with extensive choroidal detachment following trauma (patient struck head on right temple while cleaning); cataract extraction four years earlier.

Illustrating the Transparent: How to Draw the Invisible

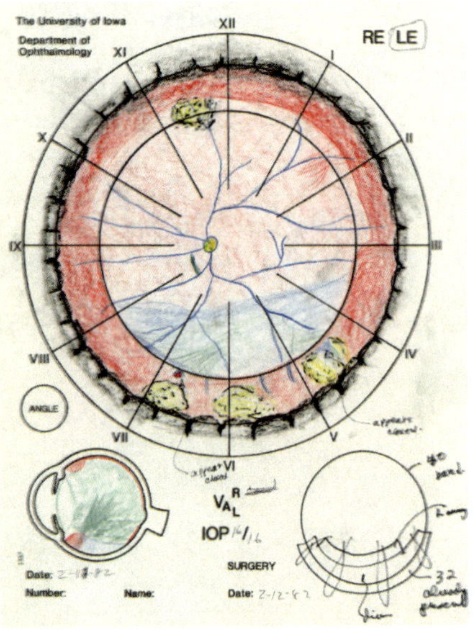

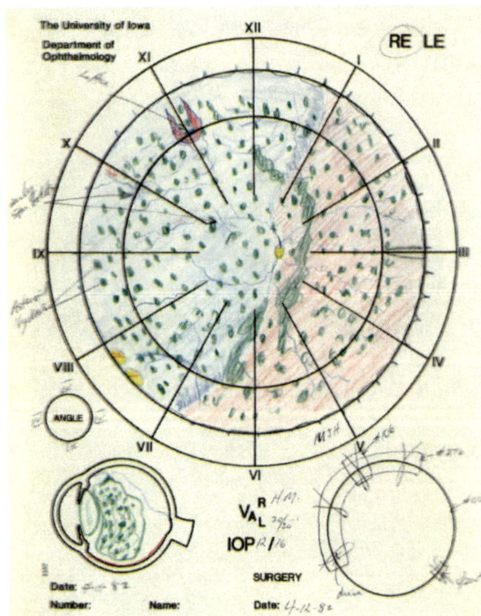

Vitreous Hemorrhages, Precipitates, and Other Opacities

Illustrating Membranous Sheets

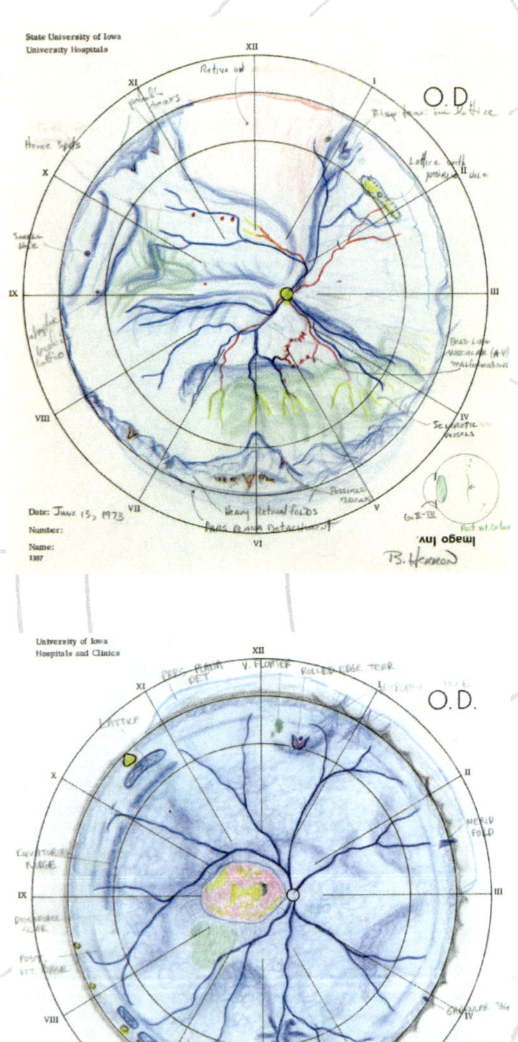

Misuse of Green: Pushing the Boundaries Artistically

Chapter 5

Limitations of the Color Convention and Use of Pseudo-Color (The Green Chapter)

Within the spectral range of the structures and abnormalities seen in the retina, there are no green hues.

Red (from hemoglobin), yellow (from xanthophyll), and black (from melanin) are the dominant colors and pigments within the fundus. Occasionally blood (hemoglobin) during its degradation process will appear relatively green. Green was designated by the color standard to represent opacities within the vitreous and other media (lens, cornea, anterior chamber). The most frequent material that opacifies the vitreous is hemorrhage (blood). Ironically, red (blood) is the complementary of green (i.e. color "opposites" when the colors are displayed on a wheel).

In a wider context, green was used to represent opacities of any degree, whether subtlety such as transparent or translucent vitreous membranes, haziness within the vitreous from cellular growth in such disorders as proliferative vitreoretinopathy, or more dense or complete obscuration from vitreous cysts or deposits.

As in all artistic forms, conventions may have been violated for effect. In these instances, green was used for emphasis of non-opacified structures.

Illustrating the Transparent: How to Draw the Invisible

In its most nuanced application, green was used to represent minimally translucent membranes within the usually transparent vitreous body. In this drawing, the long, green line near the top illustrates the presence of one such membrane. Additional green marks around inferior holes represent condensation and the translucency of vitreous around a hole in lattice degeneration. Other green highlights may represent membranes or layered blood.

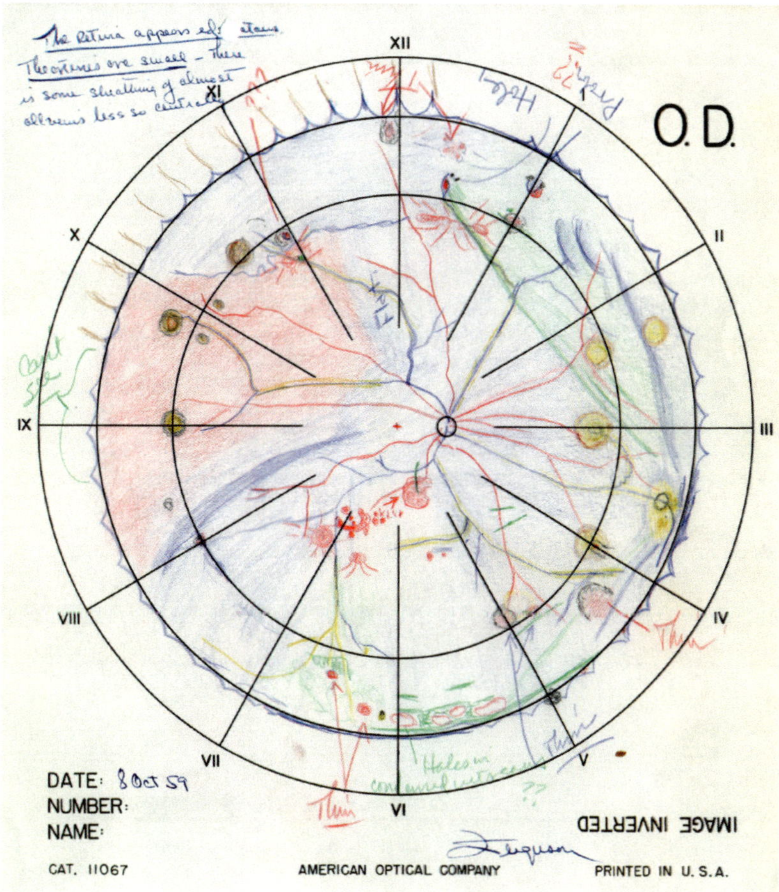

Artist Edward C. Ferguson
October 8, 1959

Diagnosis:
Recurrent retinal detachment, right eye, treated previously with retinopexy, in a patient whose left eye had been enucleated due to trauma and infection.

In the small cross-sectional figure shown in the lower left of this image from 1982, the artist illustrated that the opacification shown in green represents the formation of membranes between the inferior retinal hole and the vitreous insertion at the optic disk.

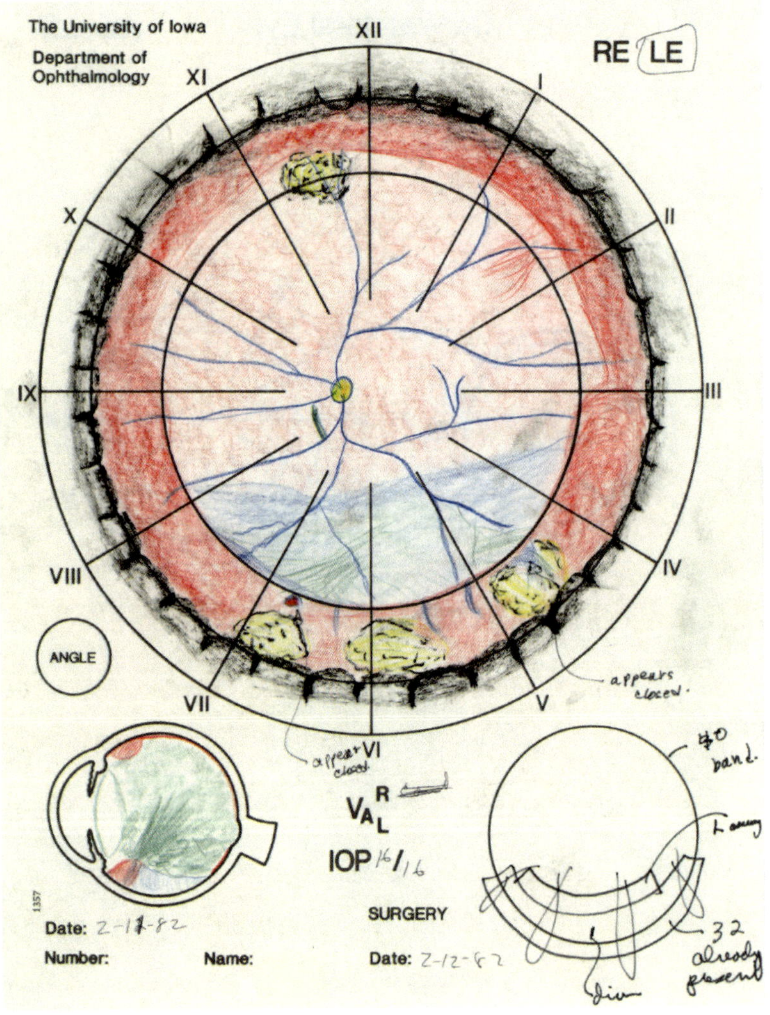

"Most of the current electronic medical records are nothing more than enhanced word processors, which do not provide a good format for depicting the pathology. A picture, e.g., an accurate retinal drawing, is still 'better than a thousand words.'"

Vern Hermsen, MD (6/26/11)

Artist Vernon M. Hermsen
February 11, 1982

Diagnosis:
Rhegmatogenous recurrent retinal detachment, left eye, previously treated, with scleral buckling.

Illustrating Membranous Sheets

Perhaps one of the most difficult challenges for the ophthalmologist was to convey the presence of clear, reflective, or minimally translucent sheets of tissue within the vitreous. In this drawing (Baller, 1965), graceful curvilinear green lines arc from the organizing and mildly translucent inferior vitreous. These reinforcing and arcing green lines suggest a sinister force, "claw" or "hand" attempting to injure the central retina, the region that provides the highest visual acuity. In the following drawing (Herron, 1973), green in the upper left represents a vitreous membrane.

"I have always believed that the discipline of performing an accurate drawing reflects the knowledge gained from the necessary examination, the accurate representation, the resultant surgical plan, and the successful outcome."

Robert Baller, MD (12/23/10)

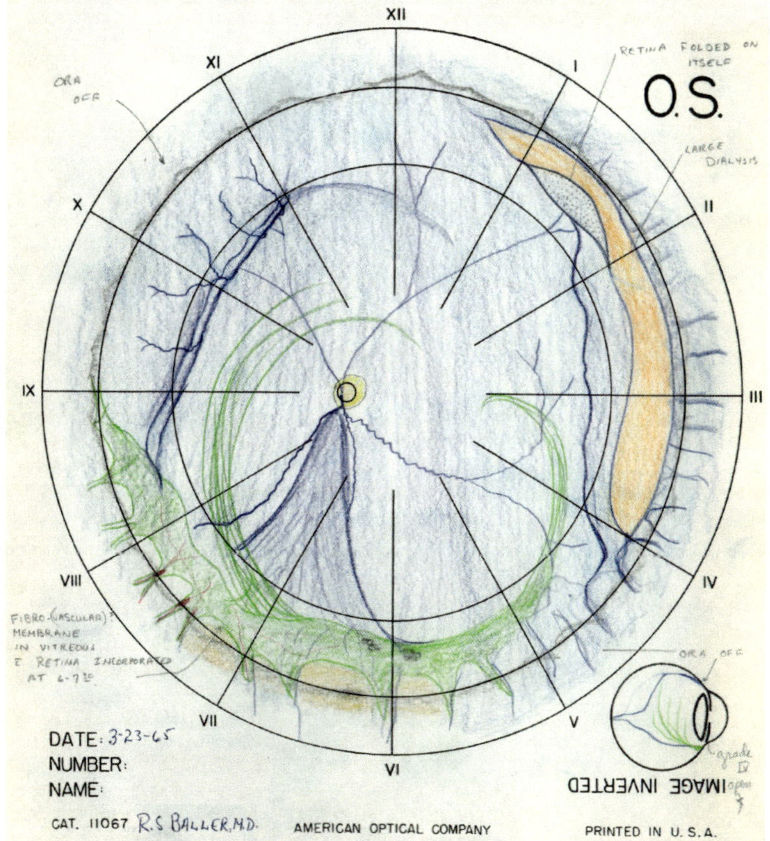

Artist Robert S. Baller
March 23, 1965

Diagnosis:
Retinal detachment, left eye.

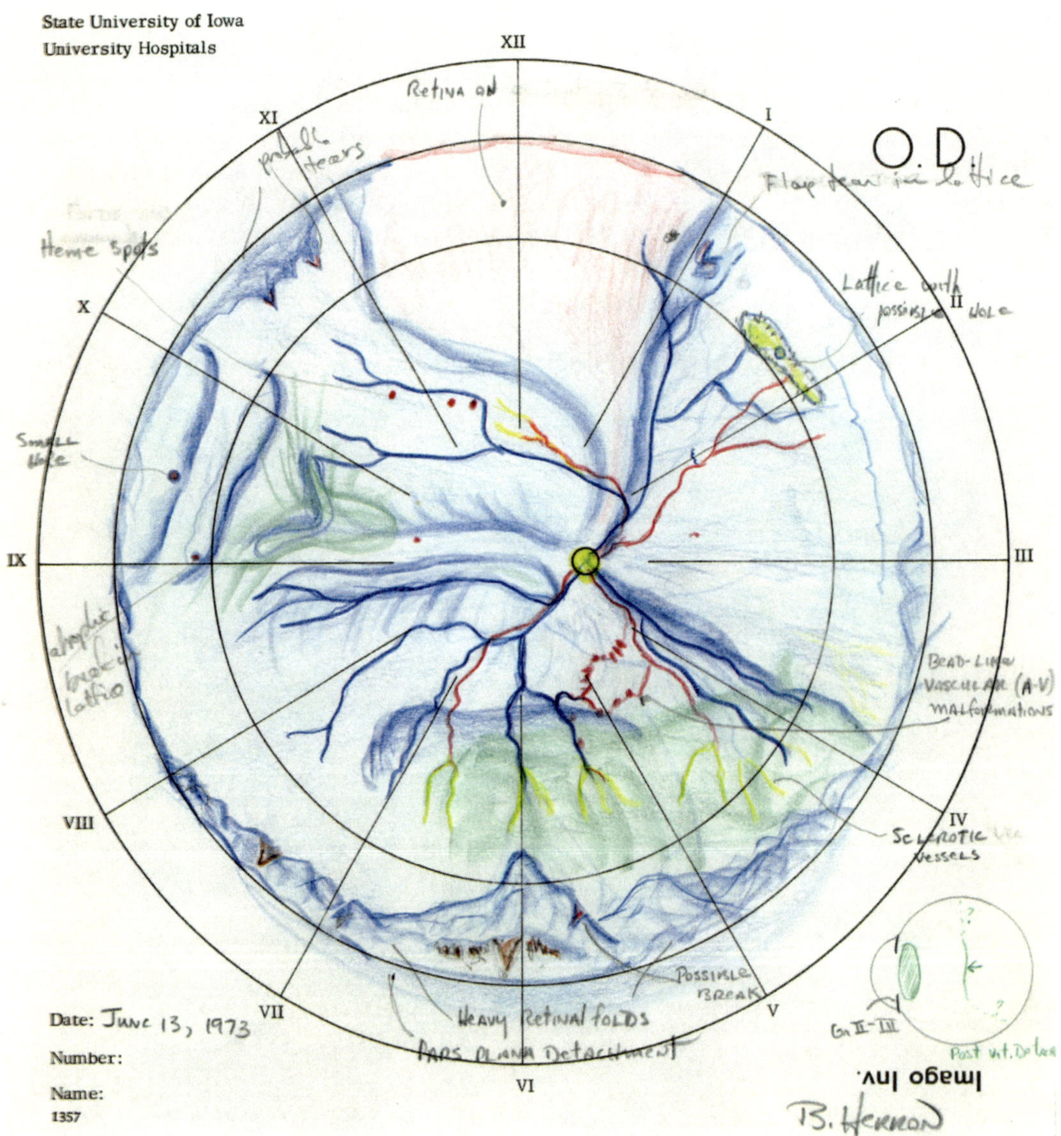

Artist Bruce E. Herron
June 13, 1973

Diagnosis:
Retinal detachment, right eye, after lifelong poor vision due to several conditions (rubeosis and chorioretinitis with optic atrophy).

"I suppose that we were shown some fundus drawings and asked to try and draw what we saw in a similar manner. I am not very creative and would have needed an example to follow. I do not claim to have much artistic talent. (I draw stick people.) The structures in the retina do not take very much skill to represent. Some are better—I guess that it depended on how much time you were willing to invest in the drawing. (Okay, some do have more talent.)"

Bruce E. Herron, MD (2/26/11)

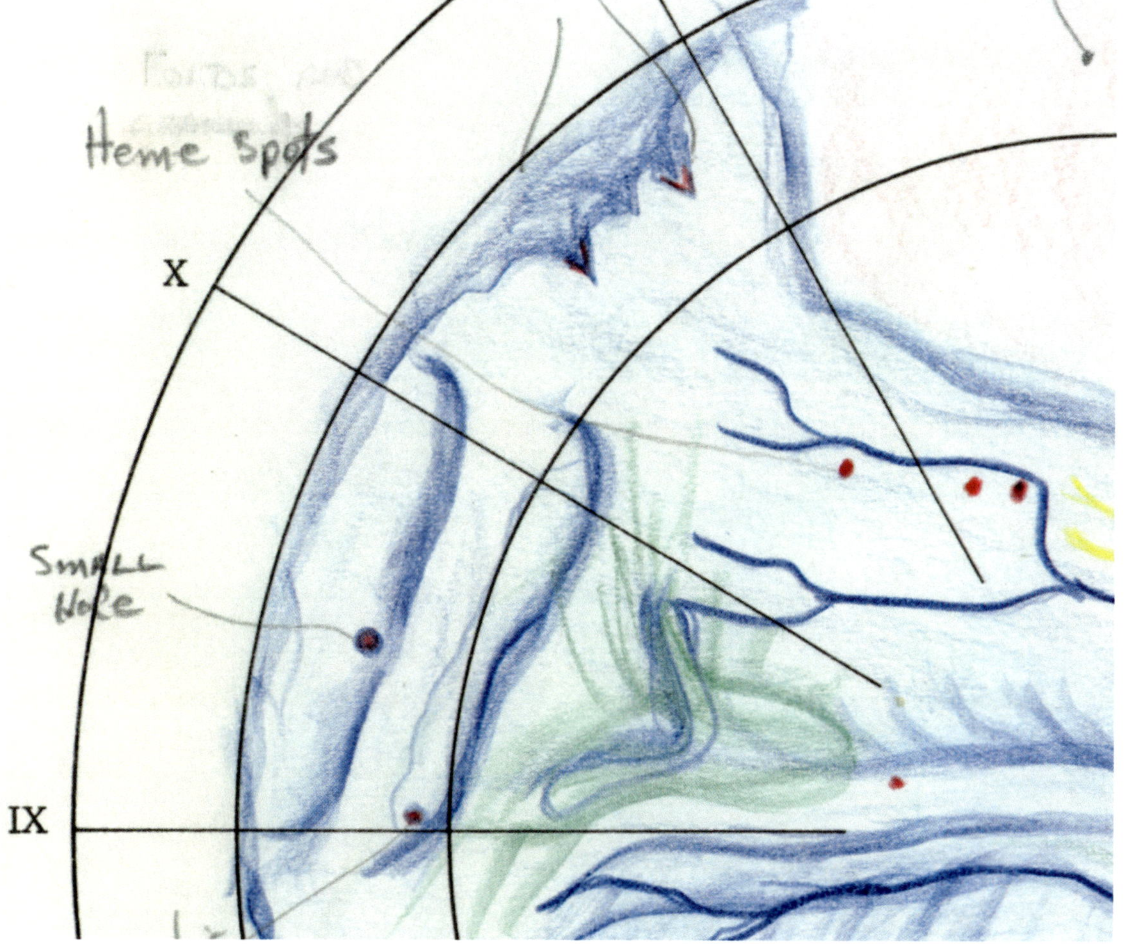

The use of green in the Caskey drawing (1987) was an attempt to illustrate the clear but reflective surface of the posterior vitreous once it had separated from the retinal surface. This surgically important surface is also known as the vitreous face.

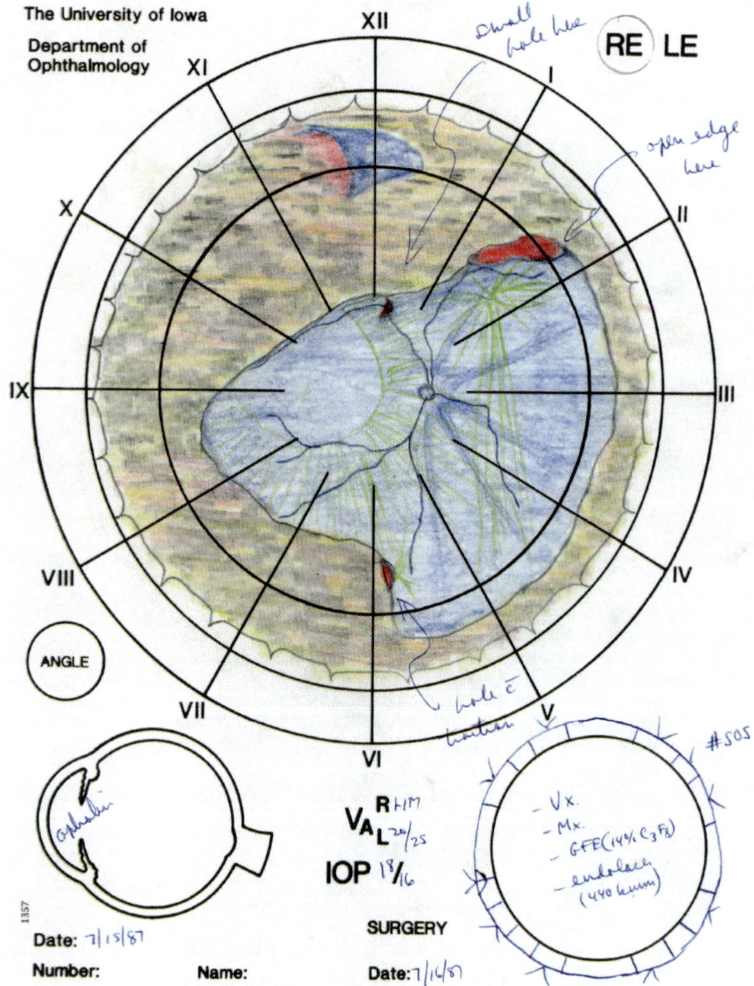

Artist Patrick J. Caskey
July 15, 1987

Diagnosis:
Recurrent retinal detachment, right eye, one year after a soccer ball hit the eye, which led to a giant tear, hyphema, and an angle recession. Previous treatments included goniosyncholysis, trabeculoplasty, trabeculectomy, and drainage of massive choroidal detachments.

Vitreous opacification, as shown in green at the bottom of the Pavan drawing (1981), obscured retinal and more peripheral details, simplifying the ophthalmologist's task of rendering. Knowing the trepidation of revealing a suboptimal drawing to faculty, there must have been a certain relief that one-third of the drawn retinal area cannot be criticized.

It is also typical of my OCD that I would draw in all the old laser spots. I am surprised I used a 41 band and a sponge. I was never a big fan of sponges or 41 bands. ... By the way, notice the annotations to drawings are in pencil. That was so Tom Burton could erase them if they were not correct."

Peter Reed Pavan, MD (12/22/10)

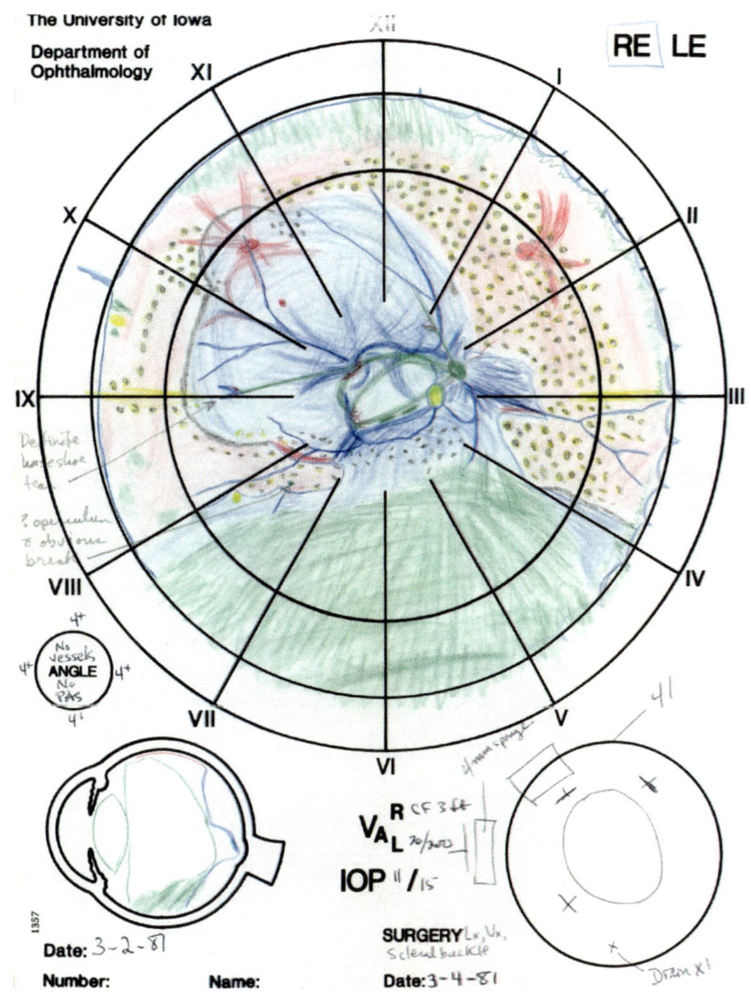

Artist Peter Reed Pavan
March 2, 1981

Diagnosis:
Recurrent tractional and rhegmatogenous retinal detachment, right eye, with a history of proliferative retinopathy and extensive vitreous hemorrhage.

Vitreous Hemorrhages, Precipitates, and Other Opacities

The structures and abnormalities captured by green images range from large, gross vitreous hemorrhage to the rarely seen faintly visible impressions of major retinal vessels on the posterior vitreous face, as in this image (Caskey, 1985). The indistinct outlines of these meandering structures, which float within the eye at a different plane of focus from the plane of regard (retina), were emphasized by the generously wide, branching green bands.

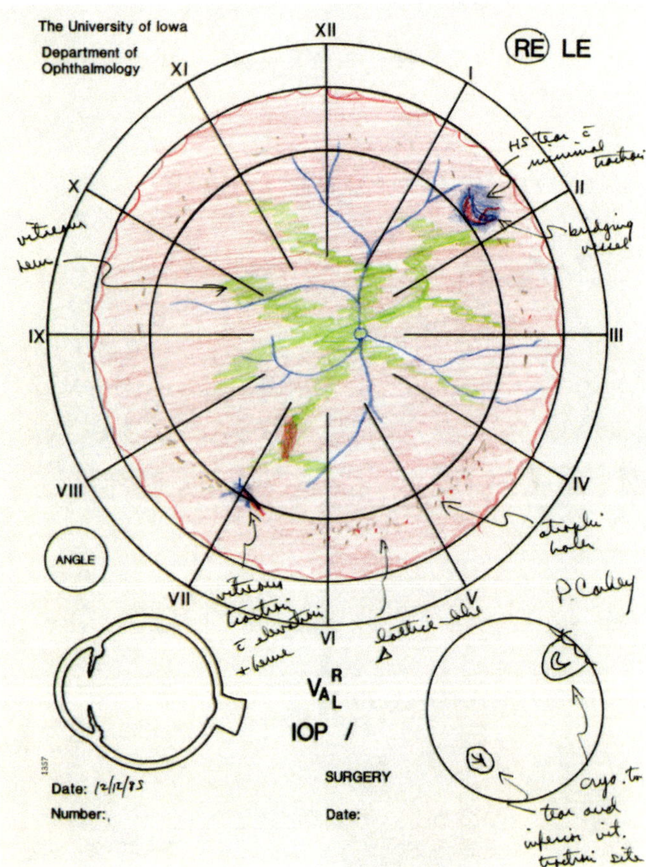

Artist Patrick J. Caskey
December 12, 1985

Diagnosis:
Horseshoe tear, right eye, with vitreous traction associated with several dot hemorrhages.

Opacification of transparent structures may be widespread or localized. Shown here is a mixture of these. In this representation of asteroid hyalosis, an incidental and harmless but visually startling and hindering multifocal vitreous calcification of unknown cause, green was used in polka-dot fashion by Dr. Howcroft to represent these dynamically moving and brightly reflective spherules that overlay a simple drawing of a retinal detachment.

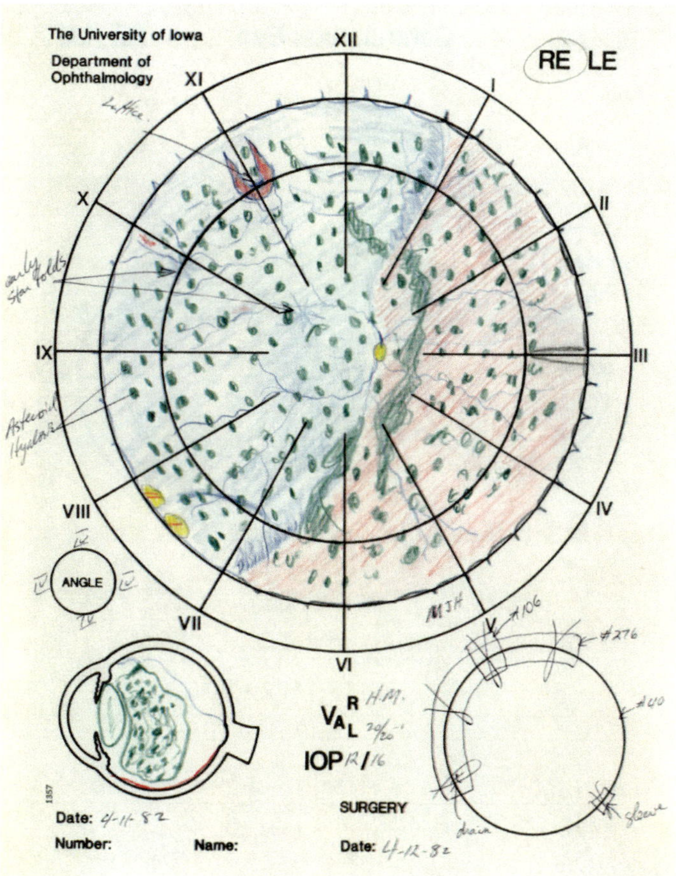

Artist Michael J. Howcroft
April 11, 1982

Diagnosis:
Temporal bullous retinal detachment and asteroid hyalosis, right eye.

The central diffuse green shading by Dr. Caskey adds a dynamic aspect to this Lichtenstein-like cartoonish drawing of peripheral cryopexy over a scleral buckle indentation (small black dots) and posterior laser photocoagulation (larger black dots).

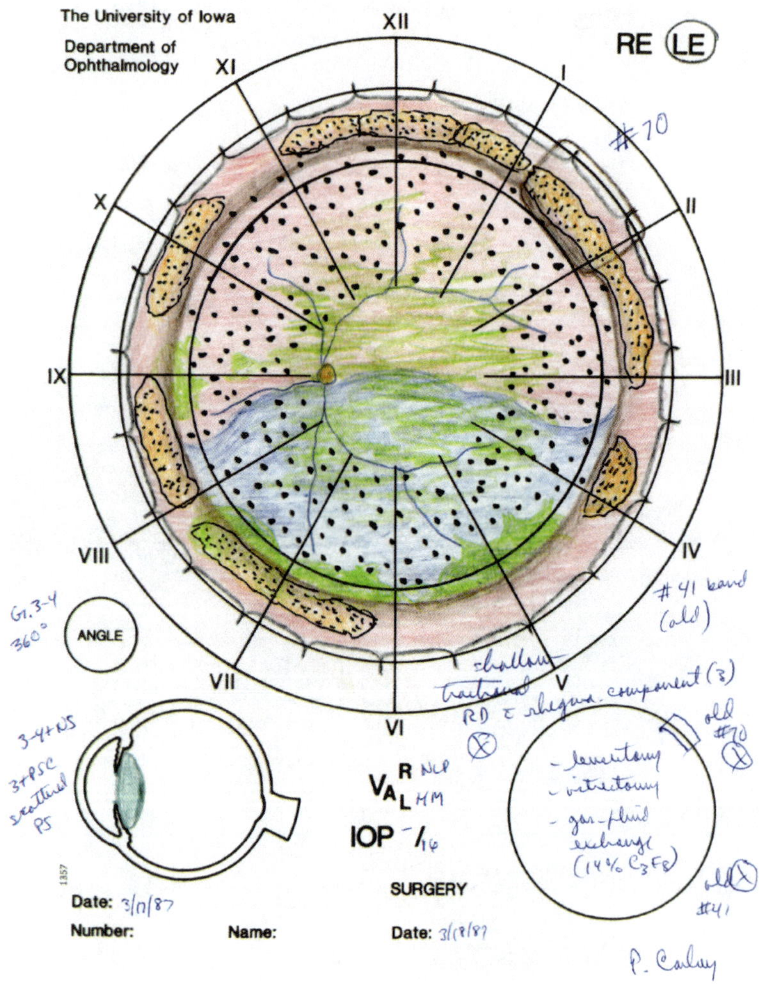

Artist Patrick J. Caskey
March 17, 1987

Diagnosis:
Recurrent tractional retinal detachment with a rhegmatogenous component, left eye, with a history of proliferative diabetic retinopathy, vitreous hemorrhage, and a cataract.

Blood or hemorrhage in the vitreous cavity is slightly denser than the vitreous body, causing the opacification to gradually settle to the bottom of the eye. Illustrated here is a typical depiction of settling vitreous hemorrhage with its variations in opacification density and its ability to partially obscure the peripheral retina.

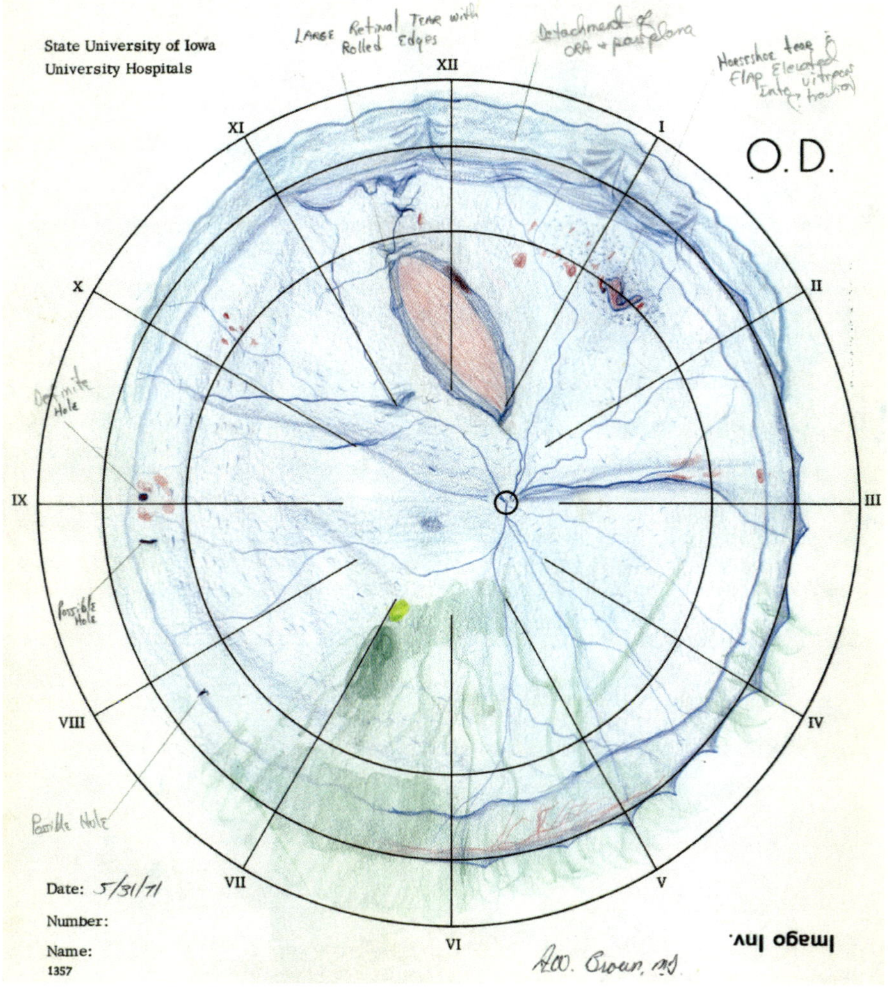

Artist Addison W. Brown, Jr.
May 31, 1971

Diagnosis:
Retinal detachment with a large tear, right eye, and Graves' disease.

Here the artist used both red and green to depict the stages or degree of degradation of a vitreous hemorrhage. Initially the hemorrhage appears red as expected. After several weeks, the hemoglobin in blood oxidizes, changing to a whitish or "more greenish" color. As mentioned previously, there is no green color or pigment in the retina or vitreous. Since red and green are complementary colors (that is, opposite on the color wheel), the loss of red appears to be greenish.

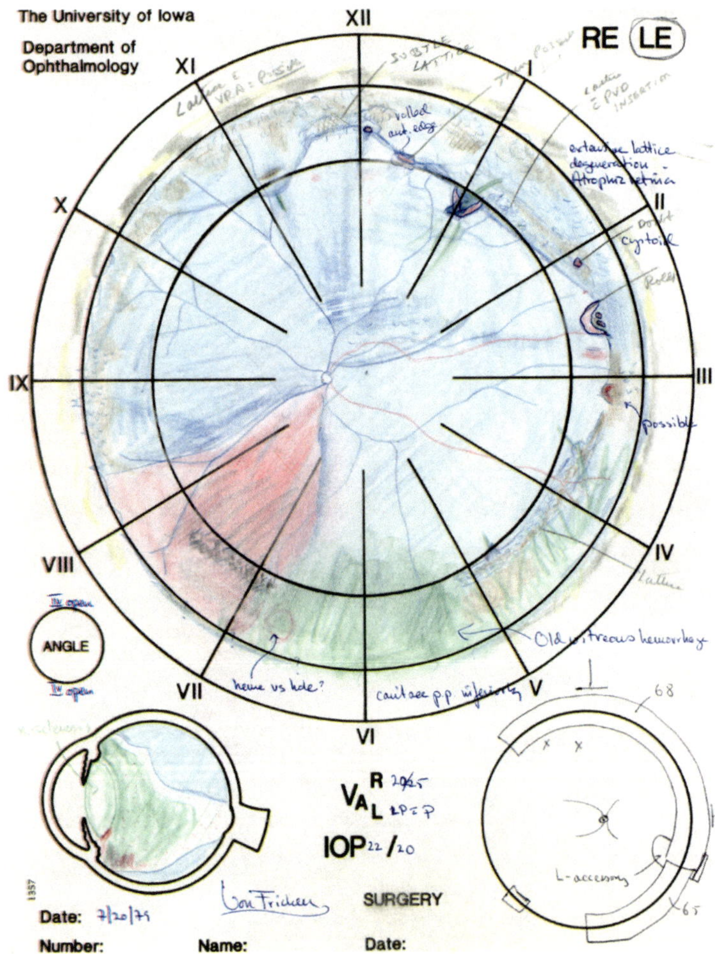

Artist Manfred A. von Fricken
July 20, 1979

Diagnosis:
Rhegmatogenous retinal detachment, left eye, with a prior vitreous hemorrhage.

The composition of this and all retinal drawings is established by the appearance of the retinal pathology. However, when critiquing the composition of this image, it must be noted that strong color blocks were used in horizontal layers. The position of red suggests fire or the sun; the blue suggests water; and the green, organic life. In the language of retinal drawings, all of these imaginings simplify to an attached retina, a detached retina, and a vitreous hemorrhage.

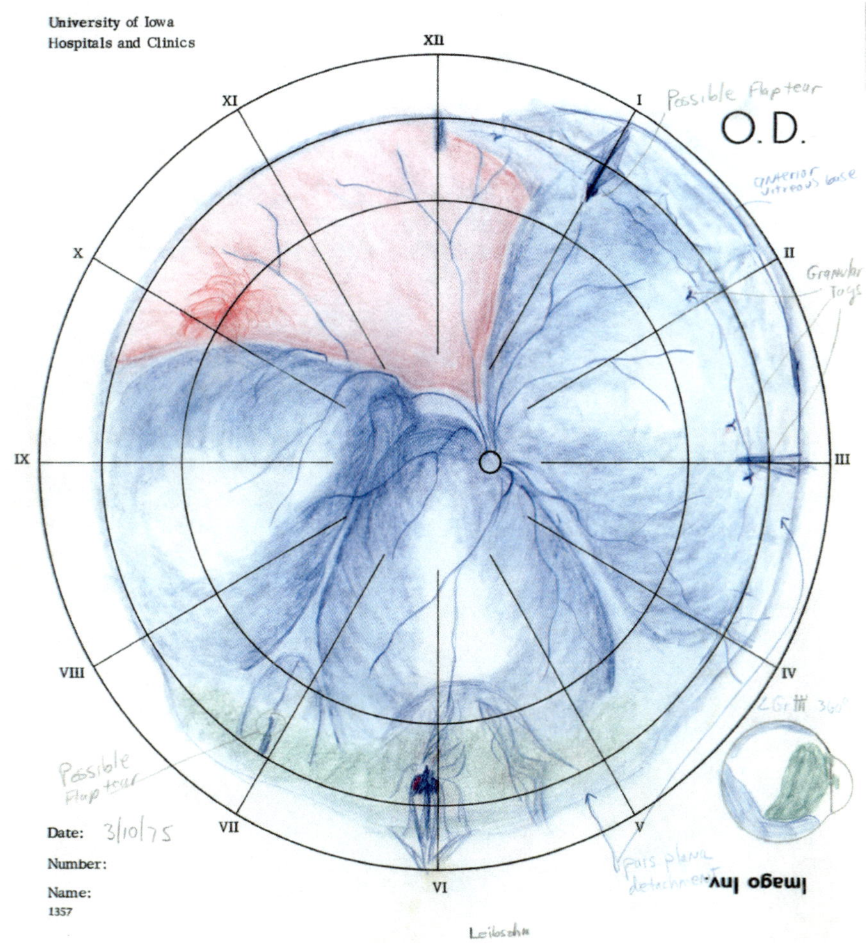

Artist Joel M. Leibsohn
March 10, 1975

Diagnosis:
Retinal detachment, right eye, six months following cataract extraction and two years following an alkali burn, also right eye.

Misuse of Green: Pushing the Boundaries Artistically

The assignment of the colors in the color code may seem arbitrary, but as was discussed previously, green may have been chosen because it is the "anti-red" (complementary color to red). Even so, given sufficient numbers of artists and enough time, deviation from the convention was inevitable. In this lovely pastel drawing, green and blue mix in the representation of detached retina. The lack of lines indicating retinal contours gives this image a still quality reminiscent of a pond in a simplistic Monet.

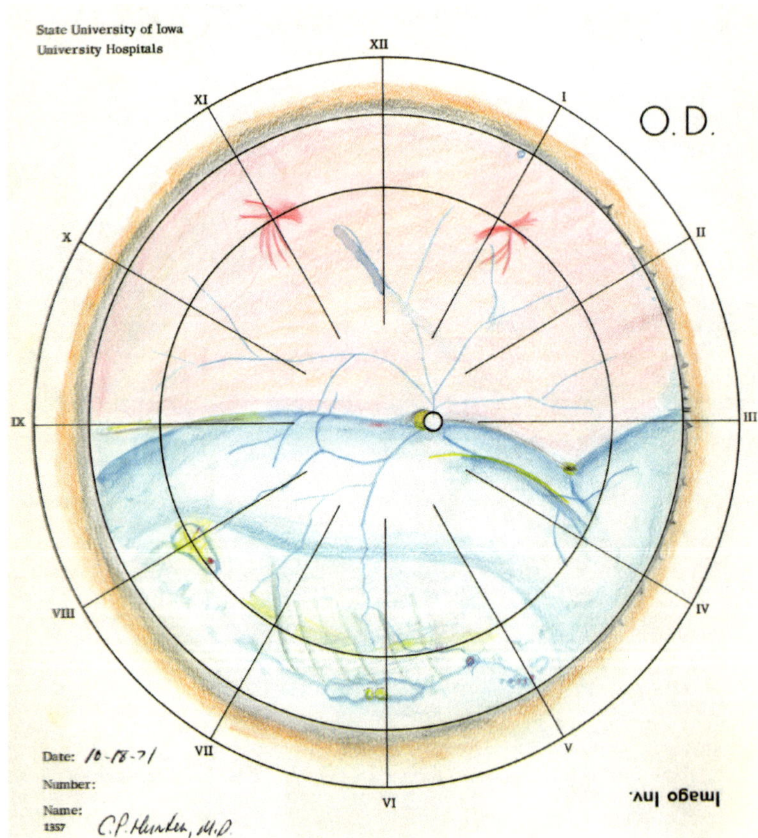

Artist Chase P. Hunter
October 18, 1971

Diagnosis:
Retinal detachment, right eye, two years after suffering a black eye during a fist fight; lattice degeneration, both eyes.

Sometimes interpretation of a combination of features allows comprehension of what the artist is trying to convey. In this drawing, there is a (green) vitreous opacity of approximately the same size and ovoid shape as the red macular hole. A beautifully drawn flap tear is present at 12:30 with a small round hole in the flap. The red island in the sea of blue is precisely the reverse of reality. The central red is a depression due to the loss of retina.

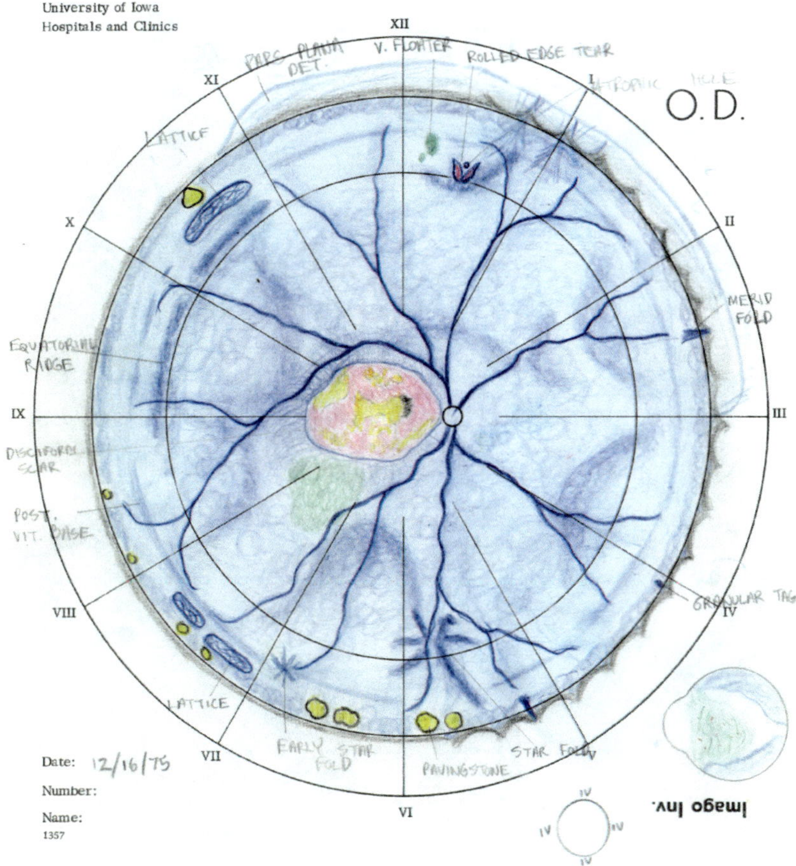

Artist Robert M. Lang
December 16, 1975

Diagnosis:
Retinal detachment, with a flap tear and lattice degeneration, right eye, following cataract extraction eleven months earlier; disciform degeneration in both eyes.

In rare instances the peripheral vitreous body becomes faintly translucent, as illustrated here, for reasons still unknown. The more significant questions are those that are often unasked. For instance, why is the vitreous clear in the first place? What physical characteristics allow it to be clear? Are not all other tissues opaque, especially those devoid of cells to dehydrate them (like the lens and cornea of the eye)? The answer is complex, but the vitreous is the only bodily tissue that is both hydrated and transparent. Given that, it may be that the clouding of the peripheral vitreous, a region that does not affect visual function, is acceptable after all.

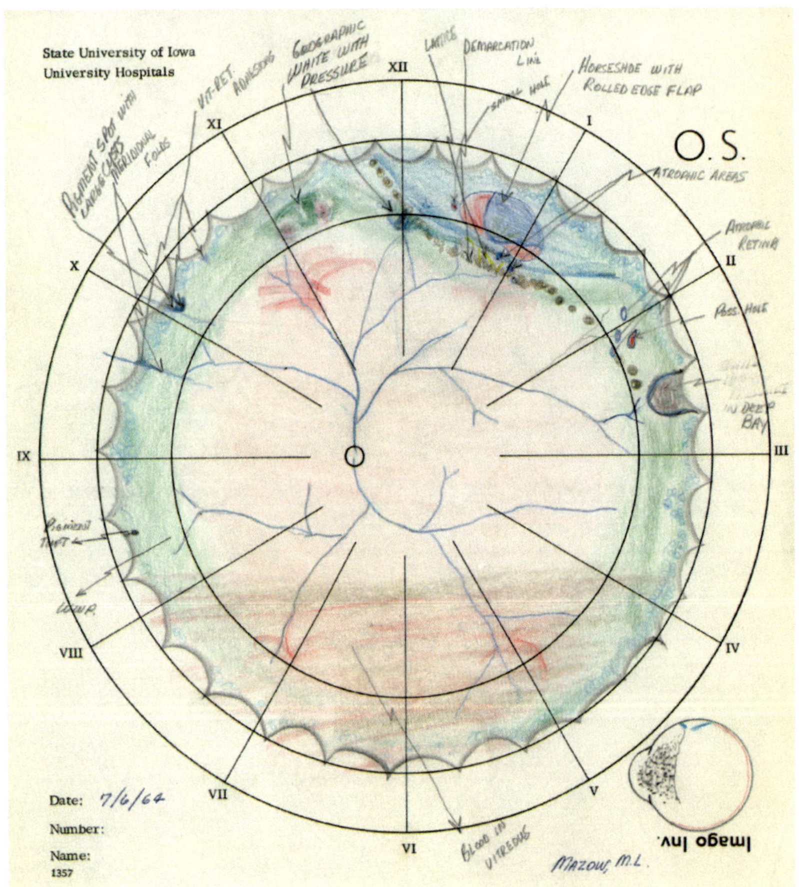

Artist Malcolm L. Mazow
July 8, 1964

Diagnosis:
Horseshoe tear with a small focal detachment, left eye, following cataract extraction four months earlier.

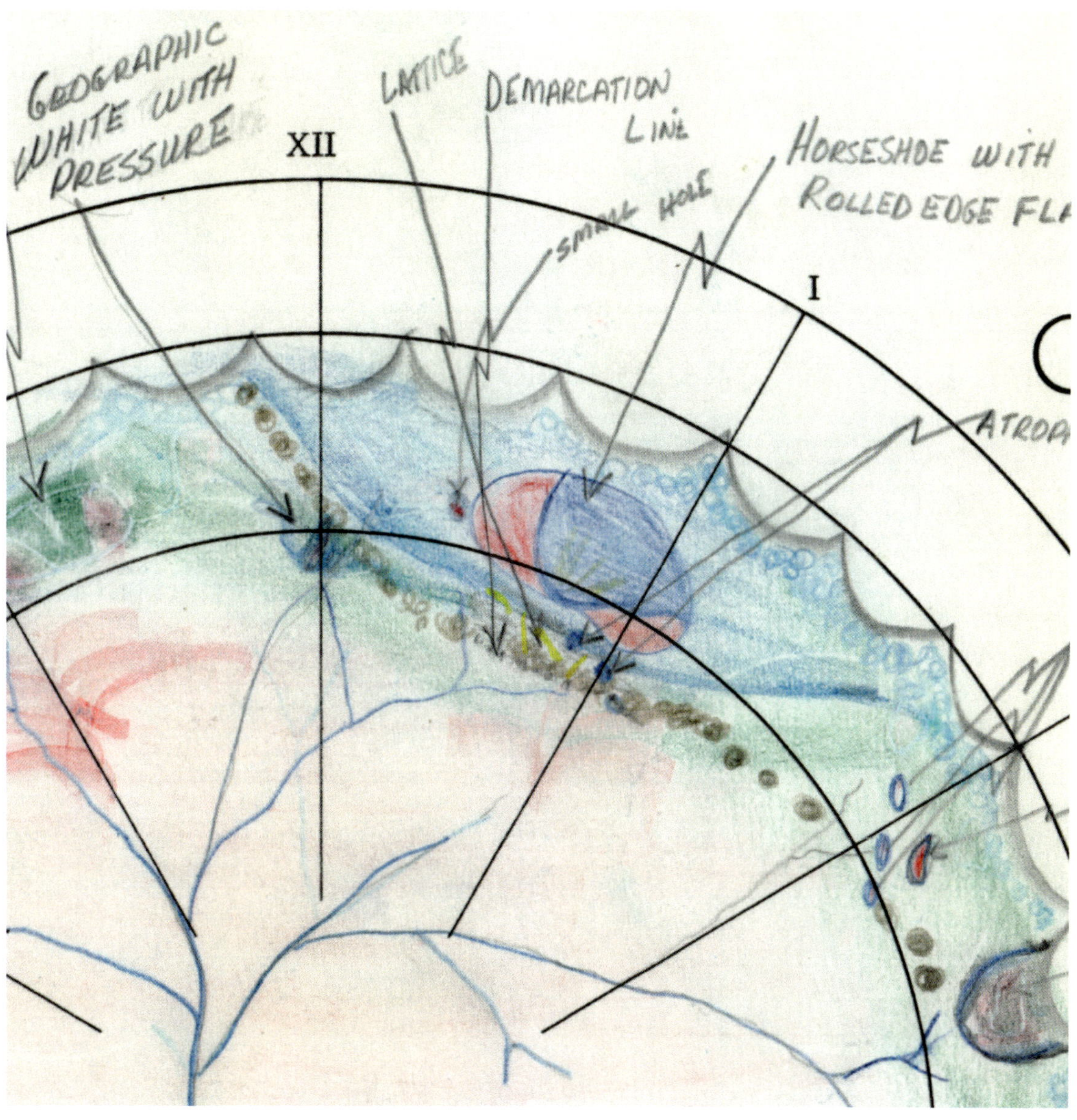
GEOGRAPHIC
WHITE WITH
PRESSURE
LATTICE
DEMARCATION
LINE
XII
HORSESHOE WITH
ROLLED EDGE FL
SMALL HOLE
I
ATROP

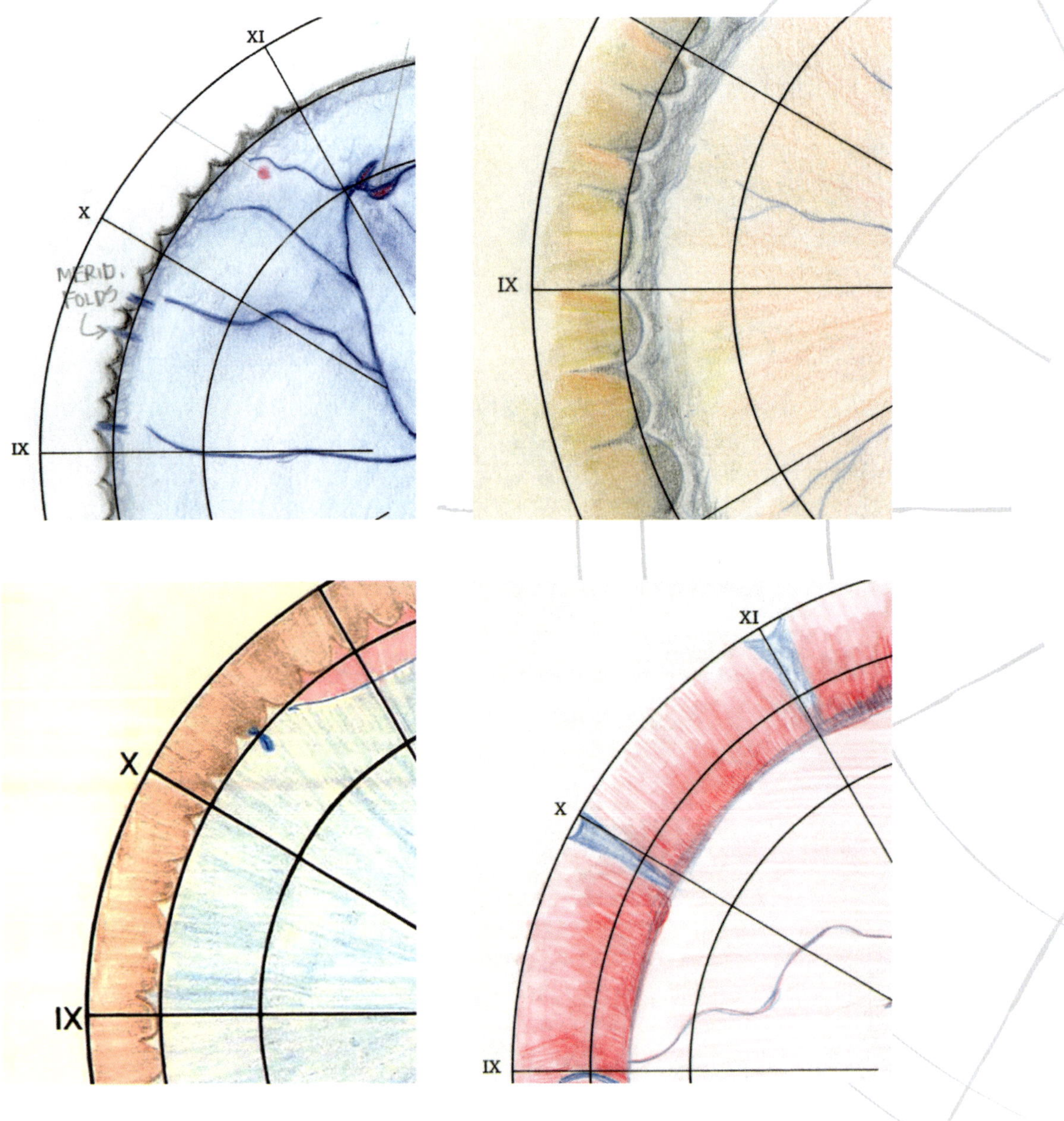
XI
X
MERID.
FOLDS
IX
IX
X
IX
XI
X
IX

Chapter 6

The Art of the Ora Serrata: The Complexities of the Hidden Edge of the Retina

During the era of formal fundus drawing at the University of Iowa, the exercise of counting "teeth," the most prominent features of the ora serrata, was rigorously enforced to the point of becoming legendary.

Among the trainees in the retina clinic during that time, it has been said and often repeated that Professor Burton, like a Marine sergeant, counted the teeth and the bays of the ora serrata on each drawing to decide whether a resident's representation was sufficient and accurate enough to be taken to the operating room. However, if that level of accuracy was so important to Dr. Burton in the 1970s, why in later drawings had these peripheral details been simplified, enhanced, or completely ignored?

On a retinal drawing, the approximate location of the ora serrata is represented by the middle of three large, concentric circles. The ora is an undulating junctional tissue ridge that designates the front or anterior edge of the retina. If the retina were a sheet, the ora would be its edge, where the retina is "tucked in." In addition to the normal junctional attachments, a layer of vitreous bridges this region, reinforcing the connective characteristics.

The ora serrata appears as a serrated edge or tissue ridge. It contains forward-pointing, thickened areas referred to as "teeth" and backward directed, rounded "bays" between the teeth. The undulations of the retinal border are more prominent on the nose, or nasal, side of the eye. Marked serration is unusual temporally or on the temple side of the eye. For a few individuals, however, a small number of small teeth may be present.

For both structural and optical reasons, visualizing the ora serrata is difficult, as it is viewed at the most extreme angle through the pupil. Because of these difficulties, photographs of the ora serrata may be difficult or impossible to obtain. Since retinal breaks in this location commonly result from separation of the retina from the anterior layers, a defect referred to as a retinal dialysis, the break can be documented by indicating the circumferential extent of the dialysis. Instead of cursory documentation, detailed drawings of the ora might seem unnecessary and time consuming.

One shortcut when drawing an ora serrata, especially when it is not fully visible, would be to indicate viewable landmarks, such as large teeth or bays, but render the remainder with minimal detail. Over the years, many shortcuts were taken by those drawing the ora, so why did some artists spend *extra* time capturing the ora's details, taking care to draw each tooth, even when the ora serrata itself does not contribute useful localizing information?

The answer may lie in the intent of those creating these images. Implicit in the care—beyond what was pragmatic—taken to convey subtle details, such as oral teeth, the examining ophthalmologist often intended to embellish, to beautify the abstraction of the natural image. For some, this intent to create an artistic work is the definition of art. Another might conceive of a definition of effect: if it appears to be art, then it is to the beholder.

The explicit purpose of the retinal drawing is for medical documentation. But the drawings collected in this volume reflect each creator's style and vision, as much as they represent the retina drawn. As demonstrated in chapter two, "How Group Drawing Styles Changed over Time," some ophthalmologists consciously attempted to draw the retina as it was seen. This tendency toward realism usually meant the inclusion of more details throughout the entire fundus rather than illustrating only the most pragmatic details, notably limited to the areas of abnormality.

Dr. Jepson chose to augment his drawings "with a touch of reality" (11/15/10). Dr. Burton "tried to use pastels to make [the retinal drawing] look more realistic" (12/14/10). In fact, he recalled that he "worked several weeks on shading and contours" to get it right. It is important to note, when discussing details and style, that both men were drawing retinas in the 1960s: Dr. Jepson arrived at Iowa in the early '60s and Dr. Burton in the late '60s.

In addition to each individual's stylistic preferences, one reason high levels of detail were included in the early retinal drawings, such as for the ora serrata, was related to the precision of the method in which retinal detachments were repaired in the early formal fundus drawing era.

Dr. Watzke stated that a related advantage for the detailed drawing was that it resulted in shortening the duration of surgery. "In the early days when exam of retinal detachments was new," he said,

> *we made a big point of drawing details of the peripheral retina and ora. We divided holes into definite, probables and possibles. The peripheral retina is so thin that when you started, you found many possible holes. In the ora, each one had to be "proven" by putting a diathermy (electromagnetic cautery) mark under it. If it was a thru and thru hole, it would look red surrounded by a white retinal burn. This was incredibly tedious and made the operation take 3–4 hours. We soon learned enough to make such decisions prior to surgery. So the early 1958 to 1970 drawings had more ora details. (4/26/11)*

This preoperative planning was why former trainees viewed fundus drawing as a map for surgery.

"At one time when most retinal detachments were corrected with only a scleral buckle," Dr. Michael J. Howcroft stated, "it was vital to have an accurate drawing for buckle placement" (6/6/11). Locating holes and identifying landmarks were part of that process.

In the mid-1970s, vitrectomy became an increasingly available surgical option and, after that, there was a transition time when preoperative planning inertia continued. Dr. Krummenacher explained that this was a time when

> *retinal detachments were managed with either a scleral buckle or a vitrectomy and the choice of intervention was made in the days prior to surgery. Unlike today, the addition of a vitrectomy was not an intra-operative option, for the vitrectomy machine was not [always or often] available. (6/6/11)*

Preoperative preparation differed between these procedures, Dr. Farmer explained:

> *There is a huge difference in the amount of conceptualization between the two techniques. Scleral buckling requires a high level of analysis and reflection. The appearance of the retina at the conclusion of surgery is incomplete...much happens after the patch is on, so a fairly thorough understanding of retinal physiology is required.*
>
> *Vitrectomy, on the other hand, is direct: the tears are seen and marked, the treatment with laser is obviously visible, the retina is pushed into place rather than coaxed, and the final appearance matches the desired post-op appearance. So there is less thinking and analysis required, and the availability of a map is unnecessary—you can see exactly where you are going, as you are working. (6/2/11)*

To more fully explain the contrast between the surgeries and why the need for detailed drawings decreased, Dr. Krummenacher added,

> *Buckling surgery mandated an accurate exam, for a missed break often resulted in a failed operation. In contrast, a vitrectomy is more of a "shotgun" approach, for after the vitreous is removed, laser is often applied to the entire peripheral retina with no adverse effects. The risk of inadvertently missing a break was nearly zero. This approach comes with its own unique risks but eliminates the risk of an incomplete preoperative exam.*

Scleral buckling, the procedure that benefited most from the meticulous retinal exam, could be supplemented by vitrectomy. And with this change, the drawings, which had already become more symbolic and less realistic, became increasingly abbreviated in scope and detail. Dr. Farmer, whose mid-'80s drawing (located in the "Iconic to Symbolic Caricature" section of the second chapter) is one of the most minimalist shown from the collection, commented, "We were not told to attempt realistic drawings, though there was general appreciation for the aesthetic value of a beautiful drawing." The appreciation remained, but the emphasis had shifted.

Overall, the drawings in this collection range from older, realistic depictions to symbolic generalizations. Even the realistic representations differ in idiosyncratic, unusual, and delightful ways from the retinal images from which they are derived. For example, in his March 1, 1963 drawing, which includes a strikingly lovely rendition of the ora serrata, Dr. Jepson drew oral teeth around the circumference in pie-crust indentations fashion; however, temporal oral teeth are rarely present and when found do not appear prominently.

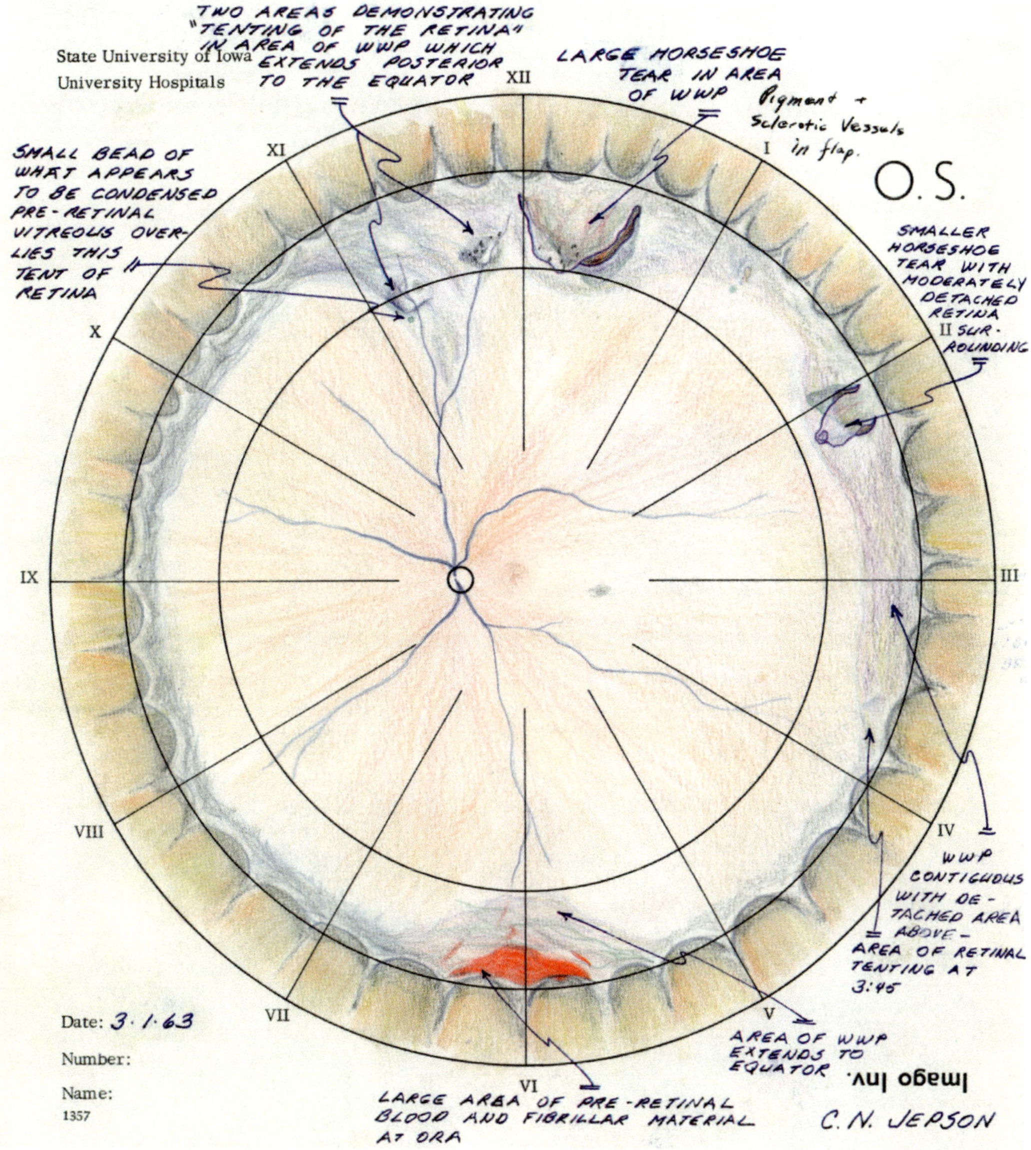

Artist C. Neal Jepson
March 1, 1963

Diagnosis:
Tractional detachment with three open tears, left eye, following cataract extraction two months earlier.

Dr. Edwin M. Stone's drawing has similar stylistic alterations. The ora appears uniformly rendered around the retinal circumference. Nasally and temporally, the oral teeth and bays are similar in size and distribution. In another stylistic alteration, the artist did not use black, the color designated by the color code, to represent the ora serrata. Apart from this deviation, however, the drawing conforms to the color code.

Also in Dr. Stone's drawing, there appear to be two major blocks of color, as could be found in many drawings from the 1970s. However, the colors present in this image are more muted than in the '70s, and the image is simpler in form. Dr. Stone's drawing harmonizes with others in his era, the 1980s, with a tendency toward artificiality and symbolism. His drawing shares striking similarity to those done by Drs. Coonan and Caskey, given the uniform pastel background, both unadorned and gothically one-dimensional despite the presence of radial lines consistent with the anticipated stylistic progression.

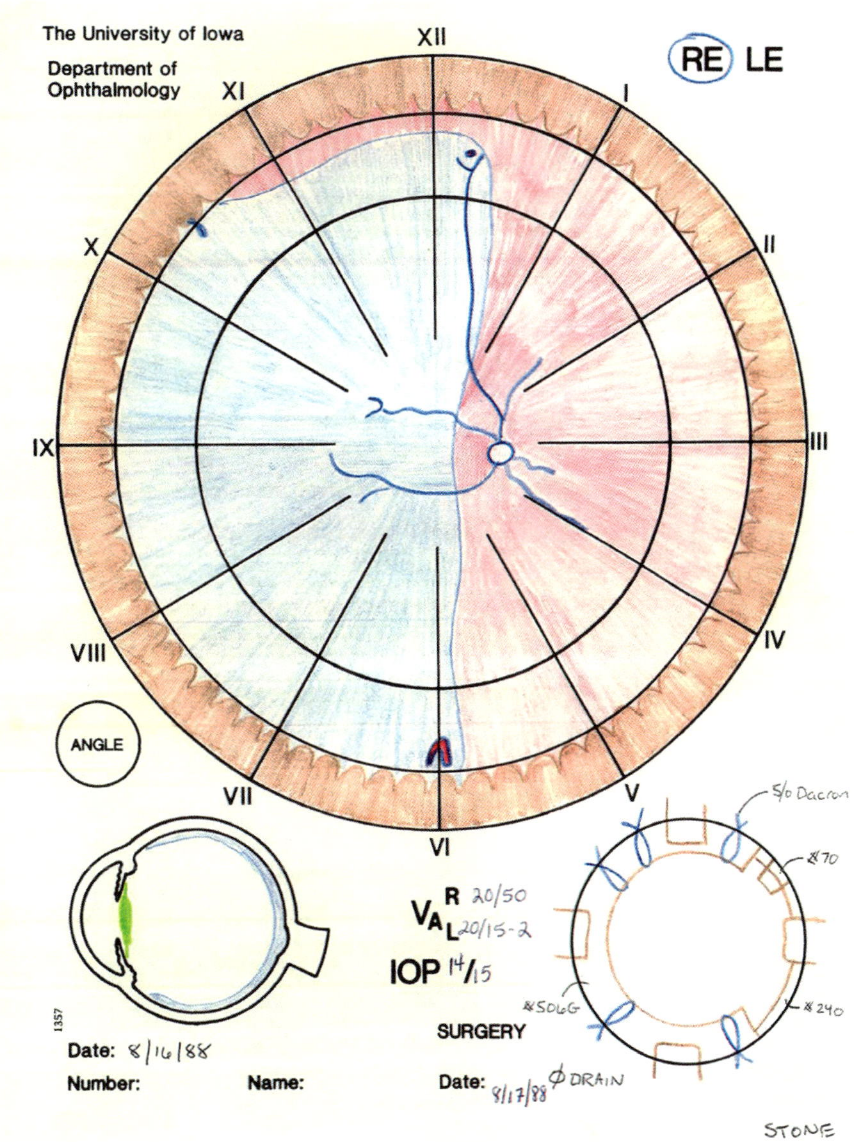

Artist Edwin M. Stone
August 16, 1988

Diagnosis:
Retinal detachment, right eye, following cataract extraction nineteen months earlier.

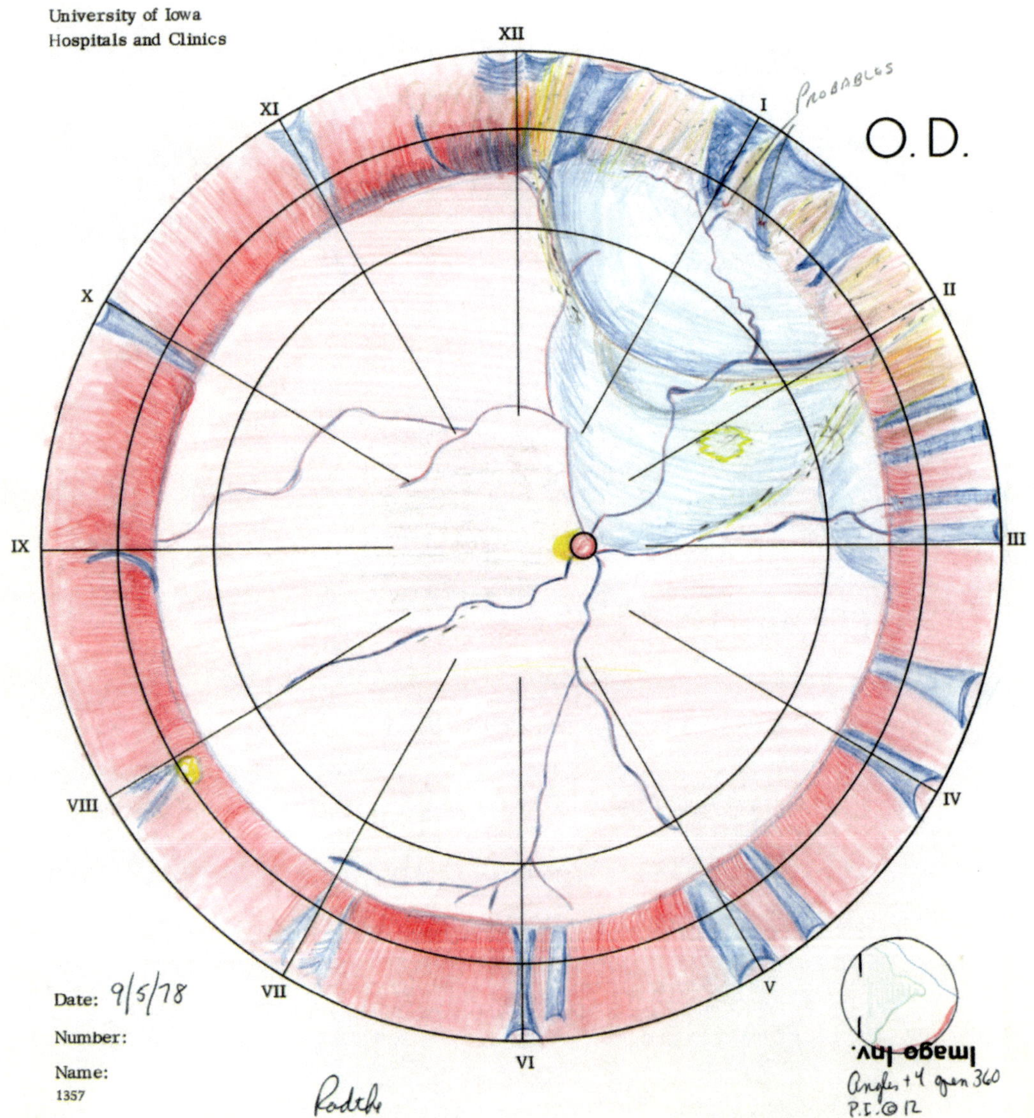

Artist Norman D. Radtke
September 5, 1978

Diagnosis:
Reaccumulation of fluid, right eye, following a retinal detachment with scleral buckle six weeks earlier and laser demarcation of subretinal fluid three weeks earlier.

Dr. Norman D. Radtke is one of three artists featured in this collection who used a distinctly brighter hue or saturation red to represent retina when it overlies a scleral buckle. (Others include Dr. Dreyer, see page 73, and Dr. Hermsen, see page 158.) Dr. Radtke's drawing from 1978 is notable because of the unusually prominent blue meridional folds, which appear to gather up anterior retinal tissue like pleats of fabric. The drawing is also noteworthy because of the shading done with curvilinear lines to represent contour or shape. This striking visual effect utilizes a different optical illusion to convey contour.

When explaining how his drawing fit into the progression of stages of the fundus drawing era, Dr. Radtke said, "The symbol of the buckle with the color to show attached retina [and] with shading to give the impression of a curve was a combination of both symbolism and realism" (8/18/11). For clarification, he added,

> *The detachment varies from bullous to flat and was depicted realistically in location but symbolically with different shadings of the colored pencils to show elevation and curvature. The shading and more enhanced areas showed the topography of the inside of the eye and was symbolic of what was going on in there but realistic as to its location as to emphasize the problem areas.*

When questioned about his artistic enhancement, Dr. Radtke continued:

> *There is no question that we enhanced the areas that we felt were the most important, such as the area of the detachment and where the possible tears were. We downplayed the areas in other quadrants that were not that vital for solving our surgical problem. ... The sections that were not pertinent to our needs were certainly downplayed and were not as accurately drawn or in the detail that we had in those that concerned us.*

Unlike a photograph that documents data, a drawing such as this one by Dr. Radtke incorporates personal interpretation. Therefore, editorializing in the form of drawing includes both information and interpretation.

In Dr. Thomas W. Smith's predominantly pink drawing from 1978, the style appears to precede similar stylistic drawings, with Dr. Smith particularly attentive to pathology and following a minimalist strategy. Comparable drawings were more often completed in the 1980s, but a few were completed in the 1970s. Perhaps instead of being notably groundbreaking, Dr. Smith may have been consumed by drawing the patient's other eye, the left eye, that day, which had recently been traumatized and was described by Dr. Victor Zion as "a very large fish-mouth, horseshoe-shaped tear with very rolled posterior edges" (1978). Dr. Zion wrote that those rolled edges were "an ominous prognostic sign because this is such a large break and the retina is not totally detached." Dr. Smith's drawing of the left eye contains oral teeth from nine to two o'clock and a thick, somewhat jagged band around the rest of the circle; whereas the right eye, more prominently featured here, contains only enough ora serrata to show the meridional folds.

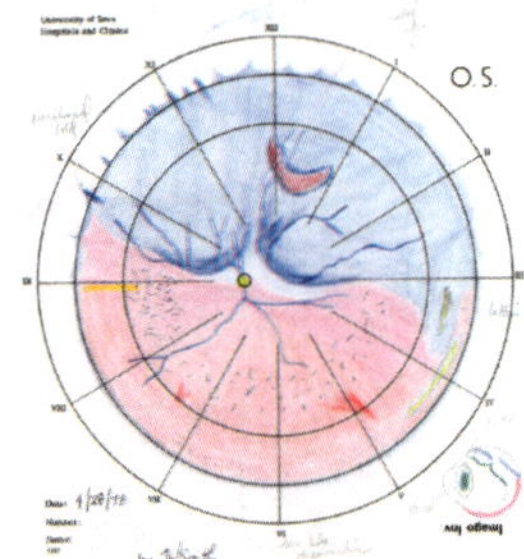

The more-detailed left eye

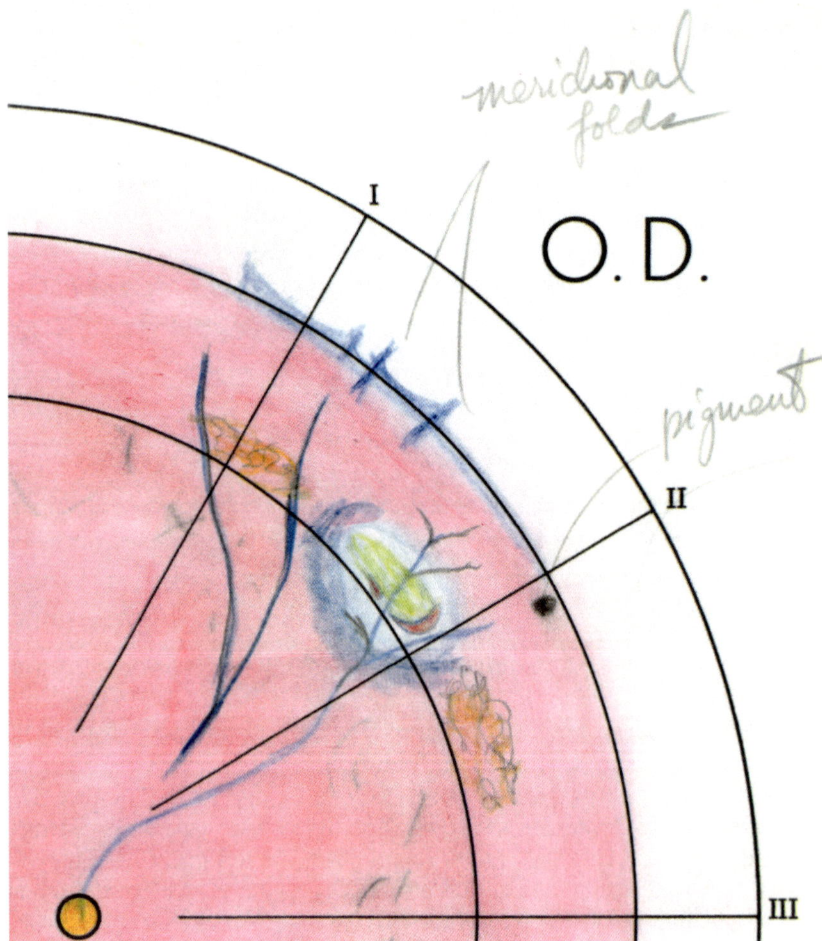

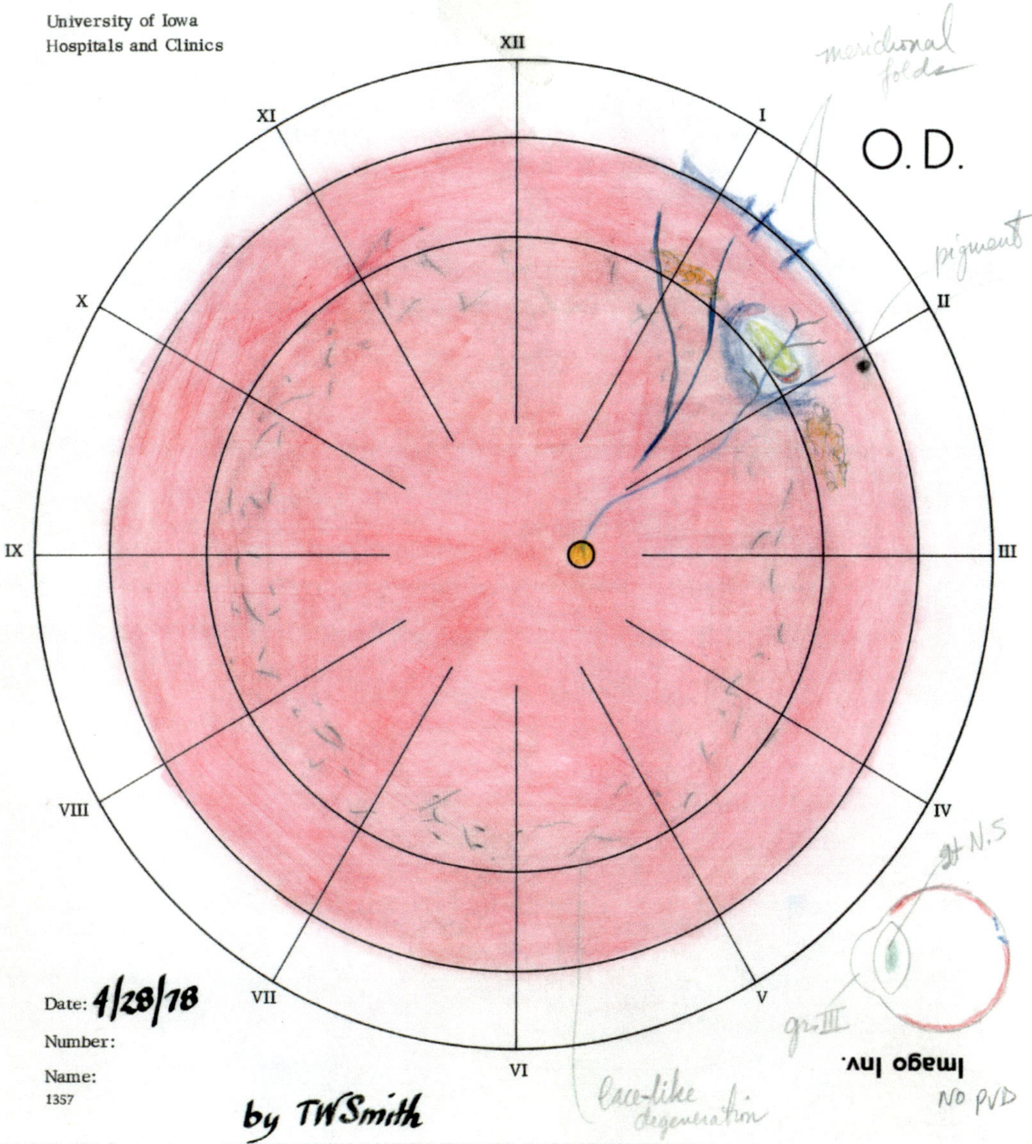

Artist Thomas W. Smith
April 28, 1978

Diagnosis:
Small pillow of fluid in the periphery, right eye.

In the 1975 drawing by Dr. Robert M. Lang, a jagged band may represent a toothless ora similar to Dr. Smith's rendering of the left eye. In Dr. Lang's representation of the ora serrata, he used black in agreement with the color code. And he drew the nasal oral teeth as though they all were approximately uniform in height, with a black band representing the temporal ora. Dr. Lang's depiction is regionally accurate, but on closer examination the reader realizes that uniformity in size and structure is rare. Variations are the rule.

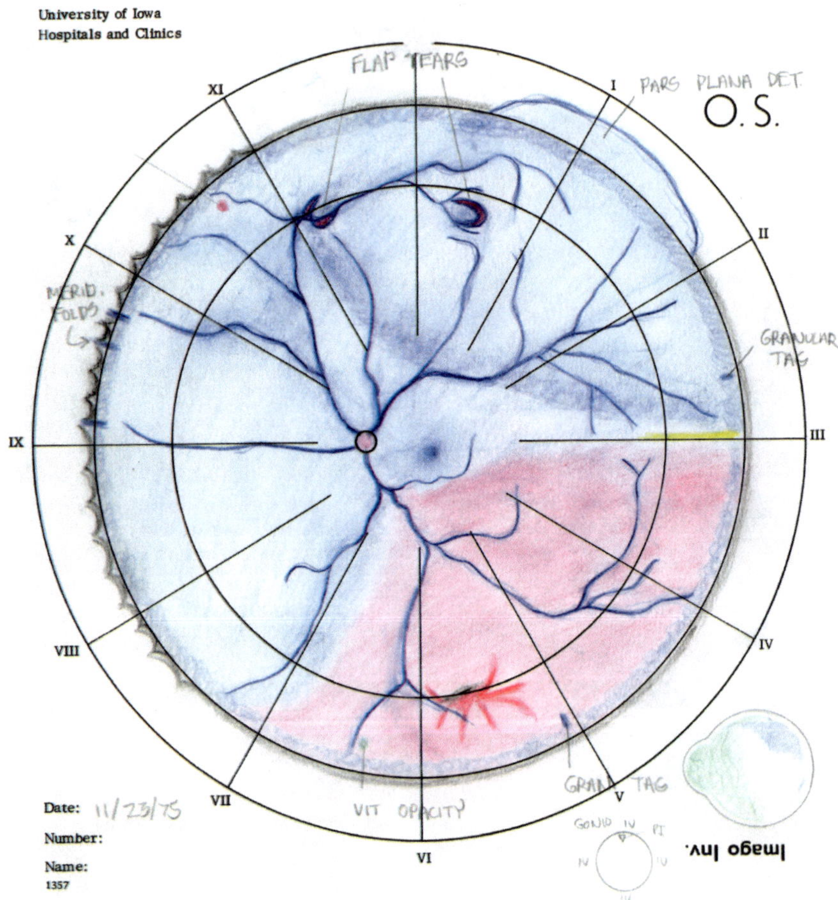

Artist Robert M. Lang
December 16, 1975

Diagnosis:
Retinal detachment, with a flap tear and lattice degeneration, right eye, following cataract extraction eleven months earlier; disciform degeneration in both eyes.

Dr. Larry W. Wood said his drawing "represented as accurate as possible detailed abnormalities observed during a comprehensive preoperative examination" (8/17/11). He added that he had attempted to "depict common differences in the ora bays and teeth between the nasal and temporal peripheral retina." The marked serration appears larger and bolder on the nasal side; whereas thin retina and snail tracks (a condition believed to be prognostically similar to lattice degeneration) are not depicted according to color code. The drawing, however, is clearly annotated to identify and document the pathology.

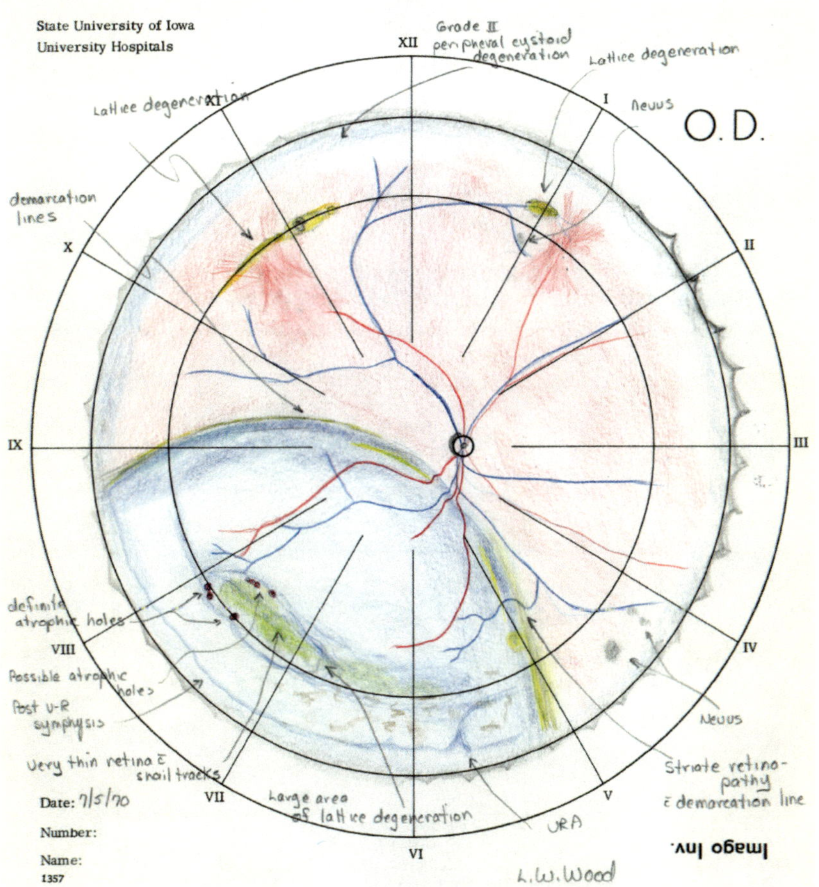

Artist Larry W. Wood
July 5, 1970

Diagnosis:
Infratemporal retinal detachment and lattice degeneration, right eye.

Retinal drawings have individual stylistic enhancements for artistic or clinical reasons. To make the drawing process more personally fulfilling or to make the picture more attractive or compelling, an examiner may have chosen to embellish rather than merely enhance, without straying so far from the color code as to confound communication of a feature's identity. Retinal drawing involved interpretation, processing, and editorial embellishment, both clinical and artistic.

An example of this is Dr. Baller's 1965 drawing. Upon first glance, his drawing appears to have elements of artistic interpretation. And Dr. Baller fully admits "some of us embellished as artists do" (8/17/11). Yet what seems to be an embellishment might just be a stylistic choice, one that another artist would not make.

Dr. Baller's annotation "ora off," for example, identifies a line signifying detached ora serrata, and yet the coloration of the oral profile changes underneath the profile line. "You can see the ora bays and teeth even though it is detached," Dr. Baller explained (8/18/11). While defending his artistic and clinical decision to draw a blue line that touches the tips of the oral teeth, he mentioned that his approach changed over time. He elaborated:

> *In the area from 9 to 12 [o'clock], that drawing embellishment was my original, signifying to me that there was peripheral vitreoretinal pathology. This was early in my career and if I saw that area again, I might handle the drawing of that area differently. … This area as I recall was assigned blue for detached retina with motivation that as such it needed attention operatively, probably by cryo now, but then probably by using scleral diathermy under flaps and extension of the buckle to flatten it out. This is an area in which mind reading comes into effect in interpreting what the artist intended to convey, rather than the artist putting down "I always mean this and it's in a form you observers can't miss." It falls into the personal note form: "Look again at this area when you (the artist now surgeon) do the surgery, and if it worries you as the case dynamically proceeds and the anatomy of the eye progressively changes toward the final result, decide if intervention there would be indicated." [It is] a nuance which is personal, in note form, not universal, where other surgeons would look at it and know the end result. (8/19/11)*

Within Dr. Baller's nuanced process and his explanation of it, there is evidence that retinal drawing evolved individually, as well as during the era.

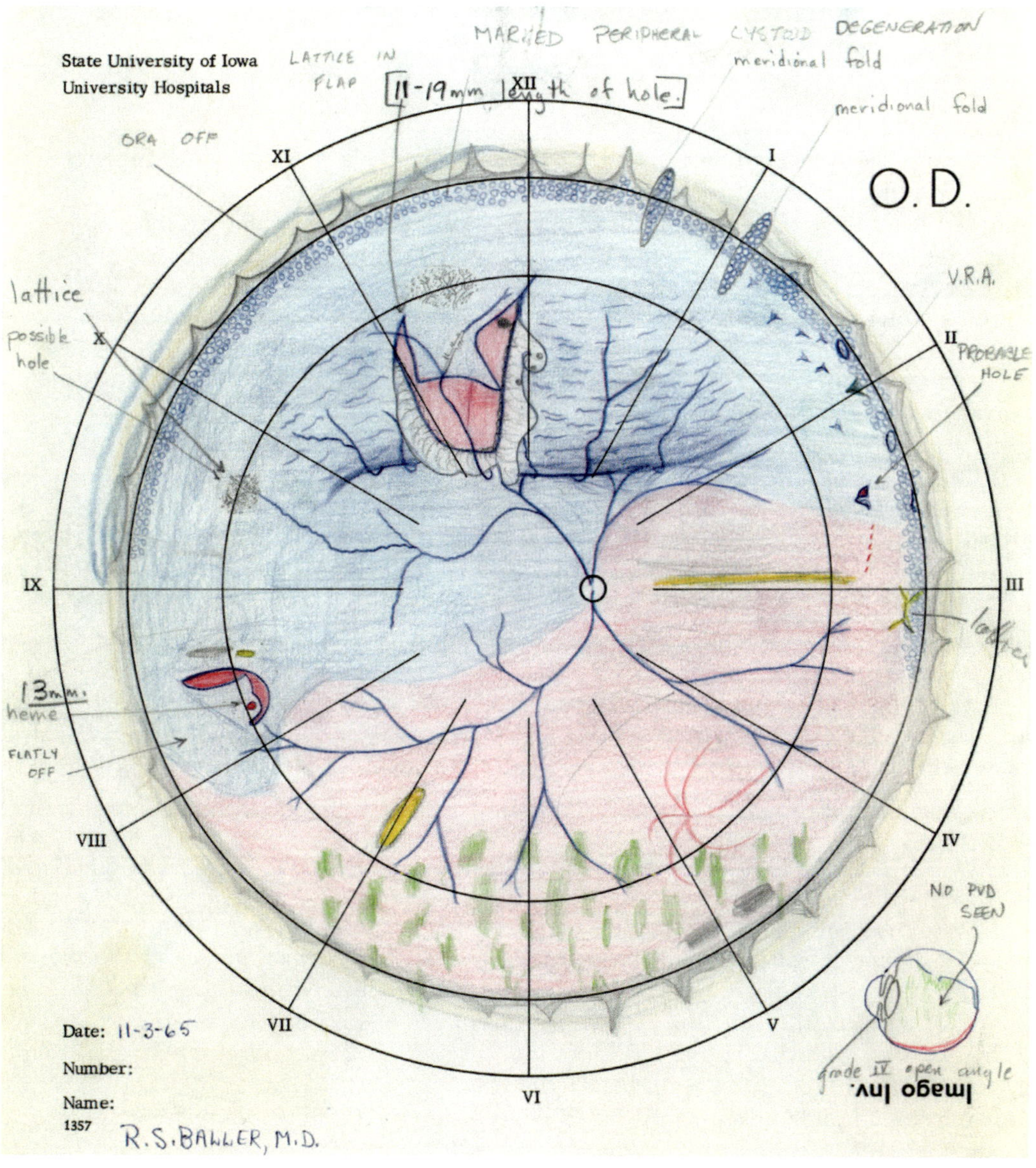

Artist Robert S. Baller
November 3, 1965

Diagnosis:
Retinal detachment with lattice degeneration and horseshoe tears, right eye, possibly related to lifting at work.

Conclusion

Although the drawings were not intended primarily to express the artist's inner feelings or primarily to influence and affect people's emotions and senses, the fundus drawings have now been removed from their original environment, just as other cultural artifacts, such as pottery, baskets, rugs, and ceremonial masks that are displayed as art in museums and galleries, have been. Those artifacts are representations of an age, a process, or an expression of individual interpretation, as are the retinal drawings. An artistic elitist might disagree with this characterization. For example, one purist has expressed the less-inclusive perspective as *l'art pour l'art* (art for art's sake). Even for the artistic elites, however, the variety and significance of these images are unique; the images represent a skill that likely will not be widely revived.

"Some contemporary artists display urinals as art," Dr. Radtke stated while discussing the "art or craft" question and its relationship to retinal drawing (8/22/11). However, the fundus drawings are "not craft," he said. "We were not making anything to sell."

The first person to display a urinal as art was French artist Marcel Duchamp, albeit under a pseudonym. Duchamp, who titled a series of everyday manufactured objects including the urinal "readymades," was said to be working against "retinal art," which is only visual. In the context of this book and the art-versus-craft debate, a demonstration against the term "retinal art" seems particularly ironic.

With his readymades and protest, Duchamp was saying that art concerns ideas, not merely what an artist sees on a table or in the distance. Of course some thought he took this concept to an extreme when at one point he placed a couple drops of paint on an amateurish painting, renamed the piece, and celebrated it as his art, since he had the idea of alteration. Formal fundus drawing involves a subtler encapsulation or conceptualization of art, a different direction than that of Duchamp, which later was seen as a precursor to "conceptual art." As *traditional* art did for centuries, fundus drawing starts with a material object, the retina, and a primary purpose.

The fundus drawings no longer hang in operating rooms; they no longer function for a practical purpose. They can be displayed and admired. And they can be celebrated for their unique imagery and as a reminder of their age. As it would be for other historical images, it may be compelling to understand why this practice was so short lived, why something that so greatly improved treatment could be discarded and nearly forgotten within fifty years. It is apparent that advancements in surgery—new procedures and new ways of treating—contributed to both the demand for and the evolution of the drawings, and also to the end of the era. Time and technological progression contributed, too, as trainees became practicing ophthalmologists and time burdens increased.

In response to why formal fundus drawing has declined, Dr. Howcroft said,

> *I think the major factor is time. These drawings often took one to two hours or more to complete (exam and drawing), and that is just not possible anymore with the pressures of daily practice. Programs like the University of Iowa, Boston (Retina Associates), and others really emphasized the drawings because it was a very good method for teaching residents and fellows how to properly examine the retina. (6/6/11)*

And yet formal fundus drawing, as it was done for three decades at UIHC, at other academic institutions, and in some private practices, ended. But without this type of sustained drawing, Dr. R. Thompson asked, "What are to become of the cognitive skills of the physician that are created by the development and complement of artistic skills?" (6/26/11)

In its halcyon days, retinal drawing was an irreplaceable teaching tool over which clinicians had discussions of lesion type, location, and association. Though time consuming for the patient and the examiner, the process resulted in unparalleled quality of instruction regarding the important visualization and recognition of diagnostic features. It is evident that teaching medical diagnosis, whether through visual pattern recognition or calculation of probability, requires repetition and careful, caring human oversight.

Within a clinical population, duplicative examination to confirm agreement introduces what some may perceive as inefficiency. There is a fundamental tradeoff between the time invested in retinal drawings and the skills and thought processes that were transferred. The process of examination duplication to some may represent time wasted but is the basis for transferring clinical knowledge. However, current health-care reimbursement schemes do not sufficiently support this kind of effort.

Dr. Diamond, who did his Iowa drawings in the 1970s, listed numerous factors, including education, time, and technology, to explain why classic drawings are no longer performed:

> *In the 80s, the Retina specialists were further aided by significant photographic advances to document these problems. It was much more desirable and efficient to take a picture than to spend the exorbitant amount of time that was needed to produce a drawing. Besides, a drawing is today more for educational development and posterity. The lack of artistic detail and attention to primarily "on spot" marking of the pathology is the intended result. Why waste time? It is not essential for treatment. In fact few individuals today draw anything other than a flat depiction of a tear, hole, or detachment in the "whimsical" fashion [that this book's authors] so adroitly noticed. (5/23/11)*

Certainly photography and other imaging technologies have been instrumental in medicine. However, some regions of the retina cannot be photographed. These areas, the forward most retinal areas and the ora serrata, can be imaged with echography or radiography or magnetic resonance (MRI) but not with direct visualization, such as photography.

Careful consideration of the situation suggests that embracing technology for technology's sake may not result in improvements in quality, but only in speed or perceived efficiency. Dr. Hermsen is one who emphasizes the multifaceted approach to retinal repair and explains why versatility is necessary:

> *Many of the conditions we now operate on, such as dense vitreous hemorrhage, do not lend themselves to drawings. The fundus is not visible or only partially visible and must be imaged with ultrasound, MRI or OCT. This has not changed the fact that careful examinations and documentation of the visible findings is still fundamental to management of pathologic retinal conditions. A picture, including retinal drawings and medical imaging, such as X-rays, CT scans, MRI, and ultrasound are the best ways to depict visible pathology. The best fundus photography cannot image well in the extreme peripheral ocular fundus. (6/27/11)*

Undeveloped skills in retinal drawing limits how well one can communicate. There are limitations, then, as well as benefits of the perceived technological improvements in ophthalmology.

At the present time, the medical system of care in the United States is moving toward less precise and specific documentation fueled by a governmentally encouraged coding and financially driven electronic format. Present medical record-keeping systems, developed primarily for non-ophthalmic care, discourage or do not allow drawings. Absent the richly edited feedback that drawings provide, it is unclear how well these visually patterned features will be recognized, understood, and taught. Dr. Hermsen laments that "the federal government and insurance companies mandate that we must use electronic medical records (EM). This has made it very difficult to document retinal pathology. 'Bean counters' love the EM because it is simple to tabulate the information they need, but the electronic medical record is a real step back for many aspects of patient care" (6/27/11).

Even now, when time and efficiency seem to be the dominant considerations, some ophthalmologists perform or teach the skill of retinal drawing in new and different ways. Dr. Diamond does black-and-white drawings that "take less than five minutes with important detail and no extras" (5/23/11). Dr. Krummenacher said he, too, still does fundus drawings. "In confession," he said, "not one of my drawings has come anywhere near the detail of my last offering in Iowa City. The point is in the end, it did not matter. It was and remains today, and tomorrow, not the drawing but the skill of the exam that leads to a better outcome for the patient" (6/6/11).

And yet Dr. Hermsen explained that in some situations more-detailed drawing is necessary, precisely because drawing engenders examination skill:

> *I still have my residents draw what they see during the fundus exam. I can quickly review the drawing and know if the student is seeing all of the pathology and recognizes both abnormal conditions and normal variations. This is very important when teaching optometry residents, because they frequently graduate from their professional school unable to recognize basic pathologic processes and normal variations. (6/27/11)*

It seems the drawings, in a smaller, often simplified, and more or less stylized form, combined with faculty review, remain an important teaching and trainee evaluation method even in a new era, and not only in university settings. Dr. Farmer, who like Dr. Krummenacher is in private practice, stated, "I believe there is an increasing sense of haste in the world, and especially in medicine, a gradually dominant business attitude that, from a desire for efficiency, may see a lovingly produced drawing as wasteful. I continue to waste time making drawings that please me" (6/6/11).

Dr. Farmer's and Dr. Hermsen's drawings include relevant detail. And, as Dr. Farmer said, his own drawings "still express the contours of the detachments via shading, and attempt to portray the specific features of the tears, their location, [and] surrounding landmarks including vasculature... features that help to express the pathology and rationalize the treatment approach" (2/3/12). Currently Medicare reimburses drawings that are closer to three instead of eight to ten inches in diameter, as long as the drawings demonstrate actual findings of specific serious pathology.

In addition to today's technological, scheduling, and political pressures, changes in other regulations and private reimbursement have led to the discouragement of retinal drawing. In the 1980s, preoperative hospitalization was increasingly disapproved as insurance carriers attempted to reduce and eliminate admissions to hospitals for surgeries. Coverage for supervision was reduced for surgical procedures as costs for and pressures against performing drawings increased. Passage of the Health Insurance Portability and Accountability Act (HIPAA) had a profound effect on the examination environment as it criminalized sharing of medical information. Prior to this act, discretion was enforced by civil law for sharing information, placing the physician at risk of a suit for financial damages or verbal constraint.

HIPAA eliminated the group examination environment so rich in educational value. No longer would retinal drawing take place in a "ballroom" or "bullpen." The drawing moved to individual examination rooms or patients' hospital rooms, where patients and examiners could no longer see other patients or hear discussions about the other patients' diagnoses. Although this shift from having multiple examinees on stretchers in one room ensured privacy, much of the educational benefit to the examiners from being in close proximity was lost. No more were colleagues and classmates "three feet away" from each other; no more were they able to share their findings and ask questions so readily, easily, or effortlessly.

Senior faculty who studied during the fundus drawing era have intimate knowledge of the fundus that future generations of general ophthalmologists and retina surgeons in training will not have. As senior faculty retire, there is a chance that the subtleties and interpretive abilities that come not only from the meticulous examination but also from the intimately interactive process surrounding it will be harmed, and the discovery of new treatments for retinal pathology will be hampered.

Index

Artists: Retinal Drawings

Artists: Additional Works

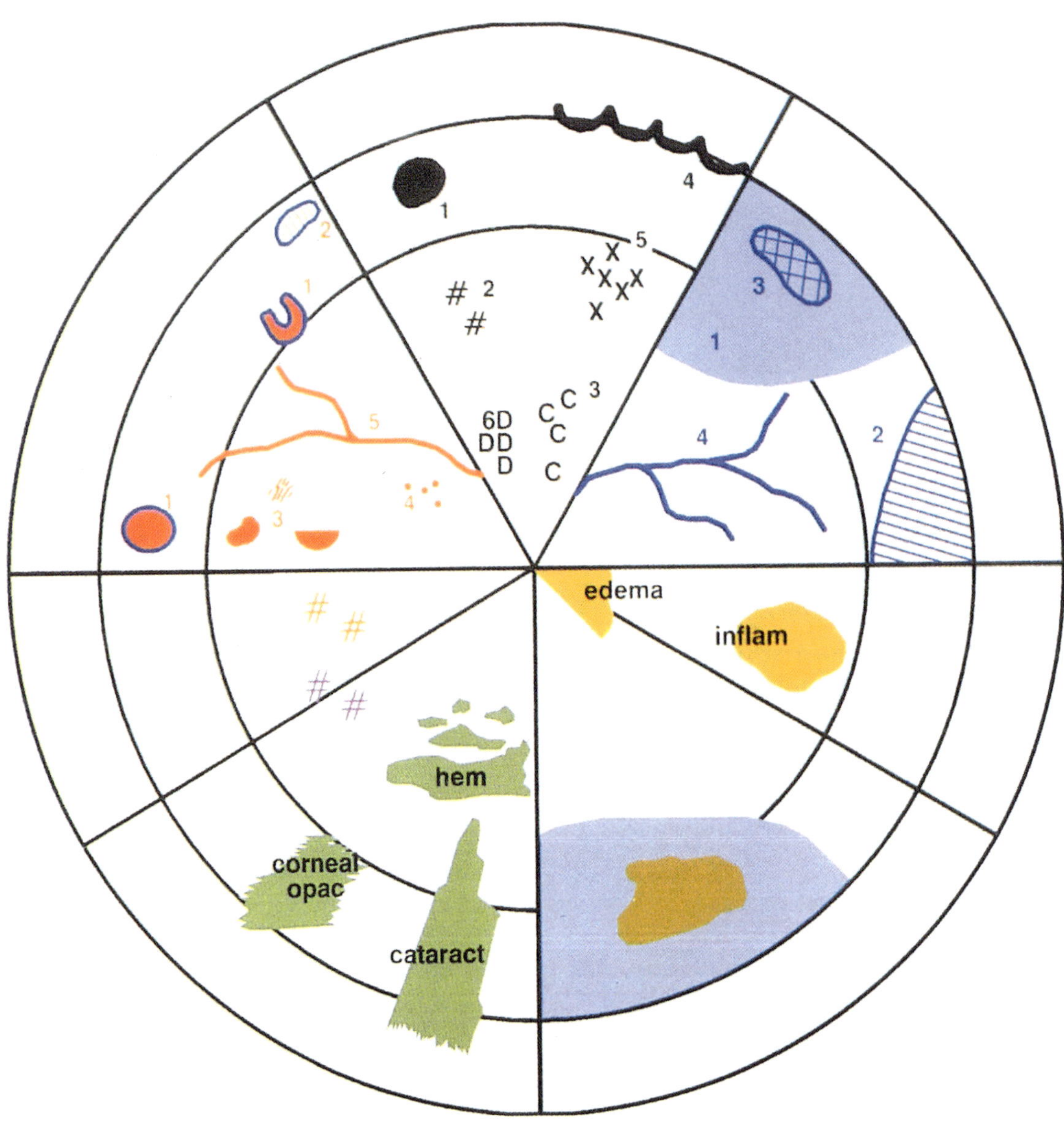

Figure 3. Fundus Drawing Color Code

Figure and table data courtesy of the American Academy of Ophthalmology with permission. Reprinted from *Retinal Detachment: Principles and Practice, 2nd Ed.* by George F. Hilton, John B. McLean, Daniel A. Brinton (1995).

Color Code for Fundus Chart

Color	Finding
Black	1 Retinal or choroidal hyperpigmentation 2 Fibrous proliferation 3 Cotton-wool patch 4 Ora serrata 5 Exudate 6 Drusen
Blue	1 Detached retina 2 Retinoschisis 3 Lattice 4 Veins
Yellow	Label specific lesions, eg, recent treatment with diathermy or photocoagulation, active chorioretinits, retinal edema
Brown	Pigment under detached retina
White	Normal attached retina
Green	Opacities of media: label specific lesion, eg, cataract, vitreous hemorrhage, or corneal opacity
Purple	Flat neovascularization
Orange	Elevated neovascularization
Red	1. Retinal breaks 2 Thin retina 3 Preretinal, intraretinal, and subretinal hemorrhages 4 Microaneurysms 5 Arteries

About the authors

Dr. Stephen Russell is Director of the Vitreo Retinal Service, Clinical Director of the Carver Center for Macular Degeneration, and the Dina J. Schrage Professor of Macular Degeneration at the University of Iowa, Department of Ophthalmology and Visual Sciences. He edited Provision 5, a compendium of ophthalmology questions and discussions for post-graduate physicians. As a retinal fellow at Iowa from 1986 to 1988, he was one of the last trainees to routinely make retinal drawings. Although he is a self-described "tech-head" (undergraduate degrees in electrical engineering and biology at Stanford University), he occasionally dabbles in writing outside of translational science.

LuAnn Dvorak is a faculty lecturer in the Department of Rhetoric at the University of Iowa, where she teaches writing and speaking to pre-health-care undergraduates. She worked for the Department of Ophthalmology and Visual Sciences at the University of Iowa during the research for and writing of this book. Her diverse background includes a newfound respect for retinal reattachment, experience in non-ophthalmic nursing, an undergraduate degree in English, and a doctorate in Education (literacy studies). She epitomizes the success that comes from interdisciplinary study. However, she most enjoys writing creative nonfiction essays about family, function and foolishness.

Made in the USA
San Bernardino, CA
04 December 2013